The F Language Guide

Springer-Verlag London Ltd.

Wilhelm Gehrke

The F Language Guide

Guide

With 79 figures

 Springer

Wilhelm Gehrke
Regional Computing Centre
University of Hannover
Hannover
Germany

ISBN 978-3-540-76165-5   ISBN 978-1-4471-0989-1 (eBook)
DOI 10.1007/978-1-4471-0989-1

British Library Cataloguing in Publication Data
Gehrke, Wilhelm, 1940-
   The F language guide
   1.FORTRAN (Computer program language)
   I.Title
   005.1'33

Library of Congress Cataloging-in-Publication Data
A catalog record for this book is available from the Library of Congress

© Springer-Verlag London 1997
Originally published by Springer-Verlag Berlin Heidelberg New York in 1997

Typesetting: Camera ready by author

34/3830-543210 Printed on acid-free paper

PREFACE

Fortran has been and will be the most important programming language for the development of engineering and scientific applications. The current standard of Fortran is Fortran 90 but its successor Fortran 95 will be accepted as international standard still in 1997. F™ is a subset of the Fortran programming language defined by Fortran 90. With very few exceptions, F is also a subset of Fortran 95. Since the "deleted language features" of Fortran 95 are not included in the F language, an F program is a Fortran 90 standard conforming program and also a Fortran 95 standard conforming program. With other words: an F conforming Fortran program is both a Fortran 90 conforming Fortran program and a Fortran 95 conforming Fortran program.

The subset of Fortran selected for the F language defines a programming language which is nearly as powerful as its parent language, which contains the modern language features of Fortran, which is significantly smaller than Fortran, which does not contain those Fortran language features that are difficult to use, to debug, and to teach, which does not contain many of the redundant language elements of Fortran, and which enforces methodical programming.

The F language does not support:

- statement labels. As a consequence, an F program cannot have or use, for instance, GOTO, arithmetic IF, alternate return, labeled do-termination statement, CONTINUE.

- storage units and storage sequences. As a consequence, an F program cannot have or use, for instance, EQUIVALENCE, COMMON, ENTRY, SEQUENCE, sequence type, sequence (argument) association.

- decremental features of Fortran 90 and Fortran 95. As a consequence, an F program cannot have or use, statement function, shared DO termination, double precision DO variable, fixed source form.

- equivalent or redundant language elements. As a consequence, an F program cannot have or use, for instance, INCLUDE, DATA, BLOCK DATA, or internal subprograms.

- rarly used language elements such as NAMELIST i/o or DO WHILE.

The name F is a trademark of Imagine1, Inc. (http://www.imagine1.com/imagine1/)

An F program requires that:

- all named data objects be declared explicitly
- attributes for data objects be specified in the type declaration statement
- user-defined F subprograms be module subprograms
- module entities have an explicitly specified accessibility attribute
- F functions have (nearly) no side-effects
- user-defined names are different from any reserved name
- the source text (except character literal constants and certain user-defined names) is written in lower case

A more detailed overview of the differences between Fortran 90 and F is given in appendix D.

"The F Language Guide" is a comprehensible description of the complete F programming language as it is defined in [1]. It is similar in style and structure to my "Fortran 90 Language Guide" [4] and is based on a revised version of that book which also was the starting text for my "Fortran 95 Language Guide" [5].

"The F Language Guide" is intended to serve as a language reference manual for the novice as well as for the experienced programmer, as teaching material for courses in F programming, and in programming methodology. The guide concentrates on the description of the language as a programmers' tool and abstains from personal, historical, and philosophical comments and interpretations.

Though the list of restrictions seems long, F retains the modern features of Fortran. Therefore, "The F Language Guide" can also be used for courses in Fortran programming. And the experienced programmer can use it during the development of new Fortran software.

Sources

[1] Imagine1, *BNF Syntax of the F Programming Language*,
     http://www.imagine1.com/imagine1/bnf.html
[2] Gehrke, *Handbuch der Programmiersprache F*, 1997, in preparation
[3] Gehrke, *Fortran 95–Nachschlagewerk*, RRZN, 1997
[4] Gehrke, *Fortran 90 Language Guide*, Springer, 1995, ISBN 3-540-19926-8
[5] Gehrke, *Fortran 95 Language Guide*, Springer, 1996, ISBN 3-540-76062-8
[6] ISO/IEC 1539:1991(E), *Fortran 90*, ISO, 1991

[3] is a German handbook similar to [5]. And [2] is the German version of "The F Language Guide".

Layout

The following conventions are used throughout this guide:

Upper-case	ABC	indicate a reserved word which must be written in lower-case within the source text.
Lower-case	abc	in a syntax rule indicate a language element which is to be inserted by the programmer.
Special characters	+ *	of the F character set must be written as given.
Square brackets	[]	in a syntax rule enclose an optional language element, which may be used or omitted.
Dot sequence	...	in a syntax rule indicate that the preceding optional language element enclosed in brackets may be repeated as necessary.
Braces	{ }	in a syntax rule enclose several language elements; one of them must be selected.

This is the font for normal text.
This is the font for formal syntax.
`This is the font for examples.`
This is the font for definitions.
Terms written such are either emphasized or are defined elsewhere in this guide.

To increase readability of this guide, keywords and other lexical tokens are printed in upper case when they appear in the text outside examples. Note that statement keywords, intrinsic subprogram names and their dummy argument names, logical literal constants and logical operators, the exponent letter in a real literal constant, edit descriptors, etc. must be written using lower-case letters when they appear in the source text of a program unit.

The formal meta language used within this guide supports the precise description of single language features. Note that it is not the formal meta language used in the original language definition [1]. Appendix E of this guide contains a representation of the formal syntax of the F language in rail-road form.

Technical Terms

The following terms are assumed as known or are defined elsewhere in this guide; in any case, reading the guide will be easier if the reader is aware of them:

block: A sequence of executable statements which is a part of an executable construct.

data object: A variable or a constant.

data entity: A data object, the result of the evaluation of an expression, or the result of a function reference.

definition: A derived type definition defines a derived data type. A subprogram definition defines a user-defined subroutine or function. A variable or a record of an internal file are defined if they have a valid value.

F processor: The computing system consisting of hardware and software by which programs are transformed for use on that system.

parent object: A subobject is a part of a parent object.

presence: An optional dummy argument is present if an actual argument is associated with it which is either a present dummy argument of the caller or which is not a dummy argument of the caller.

reference: A "data object reference" is the appearance of the data object name or data object designator where its value is required during program execution. A "subprogram reference" is the appearance of the subprogram name, of an operator symbol, or of the assignment symbol where the execution of the subprogram is required during program execution. A "module reference" is the appearance of a module name in a USE statement.

variable: A named variable is a scalar or an array object which has a name. An unnamed variable is an array element (scalar), an array section (array), a structure component (scalar or array), or a character substring (scalar).

Acknowledgements

I wish to thank Dick Hendrickson for his help in providing technical input. And I would like to thank my wife Katrin and my daughter Meike for their support and for their patience during the preparation of this book.

Hannover
March 1997

W. G.

CONTENTS

1 SOURCE FORM   **1-1**
  1.1 Classification of F Statements 1-4
  1.2 Statement Ordering 1-5

2 TYPE CONCEPT   **2-1**
  2.1 Intrinsic Types 2-1
      2.1.1 Integer Type 2-1
      2.1.2 Real Type 2-2
      2.1.3 Complex Type 2-3
      2.1.4 Logical Type 2-3
      2.1.5 Character Type 2-4
  2.2 Derived Types 2-4
      2.2.1 Derived Type Definition 2-4
            2.2.1.1 Type Component Definition 2-6
            2.2.1.2 Private/Public Derived Types and Components 2-8
      2.2.2 Structure Objects 2-9

3 LEXICAL TOKENS   **3-1**
  3.1 Scoping Units 3-1
  3.2 Keywords 3-1
  3.3 Names 3-1
  3.4 Operators and Assignment Symbol 3-4
  3.5 Literal Constants 3-4
      3.5.1 Integer Literal Constants 3-5
      3.5.2 Real Literal Constants 3-5
      3.5.3 Complex Literal Constants 3-6
      3.5.4 Logical Literal Constants 3-6
      3.5.5 Character Literal Constants 3-7

4 DATA OBJECTS   **4-1**
  4.1 Constants 4-2
  4.2 Variables 4-3
  4.3 Scalars 4-4
      4.3.1 Character Substrings 4-4
  4.4 Arrays 4-5

4.4.1   Inner Structure of Arrays   4-7
4.5   Structure Components   4-9
4.6   Automatic Variables   4-11
4.7   Association   4-12
4.7.1   Name Association   4-12
4.7.2   Pointer Association   4-14
4.8   Definition Status   4-14

5   POINTERS   5-1
5.1   Pointer Concept   5-1
5.2   Pointer Processing   5-2
5.2.1   Creation of Pointer Targets   5-2
5.2.2   Association Status   5-3
5.2.3   Deallocation of Pointer Targets   5-4
5.2.4   Nullification of Pointer Associations   5-6

6   ARRAY PROCESSING   6-1
6.1   Array Declaration   6-1
6.1.1   Explicit-Shape Arrays   6-2
6.1.2   Assumed-Shape Arrays   6-2
6.2   Reference and Use   6-3
6.2.1   Whole Arrays   6-3
6.2.2   Array Elements   6-4
6.2.3   Array Sections   6-5
6.2.3.1   Subscript-Triplet   6-8
6.2.3.2   Vector-Subscript   6-11
6.2.3.3   Array Sections of Substrings   6-12
6.3   Memory Management and Dynamic Control   6-12
6.3.1   Automatic Arrays   6-12
6.3.2   Allocatable Arrays   6-13
6.3.3   Array Pointers   6-16
6.4   Array Constructor   6-17
6.5   Operations on Arrays   6-19
6.5.1   Array Expressions   6-19
6.5.2   Array Subprograms   6-20
6.5.3   Array Assignments   6-21

7   EXPRESSIONS   7-1
7.1   Numeric Intrinsic Expressions   7-3
7.2   Relational Intrinsic Expressions   7-7
7.2.1   Numeric Relational Intrinsic Expressions   7-8
7.2.2   Character Relational Intrinsic Expressions   7-8

7.3 Logical Intrinsic Expressions 7-9
7.4 Character Intrinsic Expressions 7-12
7.5 Defined Expressions 7-13
    7.5.1 Defined Operators and Extended Intrinsic Operators . . 7-14
        7.5.1.1 Nonextended Defined Operator 7-16
        7.5.1.2 Extended Defined Operator 7-17
        7.5.1.3 Extended Intrinsic Operator 7-17
7.6 Common Rules for Expressions 7-18
    7.6.1 Precedence of Operators 7-18
    7.6.2 Interpretation of Expressions 7-19
    7.6.3 Evaluation of Expressions 7-20
7.7 Special Expressions 7-22
    7.7.1 Constant Expressions 7-22
    7.7.2 Initialization Expressions 7-24
    7.7.3 Specification Expressions 7-25

8 ASSIGNMENTS   8-1
8.1 Intrinsic Assignment Statements 8-1
    8.1.1 Numeric Assignment Statement 8-2
    8.1.2 Logical Assignment Statement 8-3
    8.1.3 Character Assignment Statement 8-4
    8.1.4 Assignment Statement for Derived Types 8-5
8.2 Defined Assignment Statements 8-6
    8.2.1 Nonextended Defined Assignment 8-8
    8.2.2 Extended Defined Assignment 8-9
8.3 Pointer Assignment Statement 8-9
8.4 Masked Array Assignments 8-11
    8.4.1 WHERE Construct 8-11
    8.4.2 Common Rules for Masked Array Assignments 8-13

9 DECLARATIONS AND SPECIFICATIONS   9-1
9.1 Attributes 9-2
    9.1.1 ALLOCATABLE Attribute 9-3
    9.1.2 Initial Value 9-3
    9.1.3 DIMENSION Attribute 9-3
    9.1.4 INTENT Attribute 9-4
    9.1.5 OPTIONAL Attribute 9-4
    9.1.6 PARAMETER Attribute 9-5
    9.1.7 POINTER Attribute 9-6
    9.1.8 PRIVATE Attribute 9-6
    9.1.9 PUBLIC Attribute 9-6
    9.1.10 SAVE Attribute 9-7

     9.1.11 TARGET Attribute 9-8
  9.2 Type Declaration Statements 9-8
     9.2.1 INTEGER Statement 9-10
     9.2.2 REAL Statement 9-10
     9.2.3 COMPLEX Statement 9-11
     9.2.4 LOGICAL Statement 9-11
     9.2.5 CHARACTER Statement 9-12
         9.2.5.1 Length Specification 9-12
     9.2.6 TYPE Declaration Statement 9-13
  9.3 Additional Specification Statements 9-14
     9.3.1 PRIVATE Statement 9-14
     9.3.2 PUBLIC Statement 9-15
     9.3.3 IMPLICIT Statement 9-16
     9.3.4 INTRINSIC Statement 9-16

10 EXECUTION CONTROL   **10-1**
  10.1 IF Construct 10-1
     10.1.1 Simple IF Constructs 10-2
     10.1.2 Nested IF Constructs 10-5
  10.2 CASE Construct 10-6
     10.2.1 Simple CASE Constructs 10-9
  10.3 DO Construct 10-10
     10.3.1 DO Statement 10-11
     10.3.2 END DO Statement 10-11
     10.3.3 Forms of DO Constructs 10-11
     10.3.4 Execution of a DO Construct 10-12
         10.3.4.1 Additional Details about Count Loops 10-13
         10.3.4.2 Additional Details about Endless Loops 10-14
         10.3.4.3 CYCLE Statement and EXIT Statement . . . 10-15
  10.4 Nested Constructs 10-16
  10.5 STOP Statement 10-16

11 INPUT/OUTPUT   **11-1**
  11.1 Records 11-1
  11.2 Files 11-2
  11.3 File Attributes of External Files 11-2
     11.3.1 File Names 11-2
     11.3.2 Access Methods 11-3
         11.3.2.1 Sequential Access 11-3
         11.3.2.2 Direct Access 11-3
     11.3.3 Form of a File 11-5
     11.3.4 File Position 11-5

11.4 Units 11-6
11.5 Preconnected Units and Predefined Files 11-7
11.6 Input/Output Statements 11-7
    11.6.1 Input/Output Specifiers 11-8
        11.6.1.1 UNIT= Specifier 11-8
        11.6.1.2 FMT= Specifier 11-9
        11.6.1.3 REC= Specifier 11-9
        11.6.1.4 ADVANCE= Specifier 11-9
        11.6.1.5 End-of-Record Condition 11-10
        11.6.1.6 IOSTAT= Specifier 11-10
        11.6.1.7 Error Conditions 11-11
        11.6.1.8 End-of-File Condition 11-11
        11.6.1.9 SIZE= Specifier 11-11
    11.6.2 Input/Output Lists 11-12
    11.6.3 Data Transfer Statements 11-14
        11.6.3.1 Formatted Input/Output 11-16
        11.6.3.2 Unformatted Input/Output 11-18
        11.6.3.3 List-Directed Input/Output 11-20
        11.6.3.4 Internal Input/Output 11-24
        11.6.3.5 Nonadvancing Input/Output 11-27
        11.6.3.6 Printing 11-30
    11.6.4 File Status Statements 11-31
        11.6.4.1 OPEN Statement 11-31
        11.6.4.2 CLOSE Statement 11-33
        11.6.4.3 INQUIRE Statement 11-34
    11.6.5 File Positioning Statements 11-39

12 FORMATS   **12-1**
12.1 Format Specification 12-1
12.2 Interaction between Input/Output List and Format 12-2
    12.2.1 Repeat Specification, Groups of Edit Descriptors 12-3
    12.2.2 Reversion of Format Control 12-3
12.3 Edit Descriptors 12-4
    12.3.1 A Edit Descriptors 12-6
    12.3.2 Colon Edit Descriptor 12-8
    12.3.3 ES Edit Descriptors 12-8
    12.3.4 F Edit Descriptor 12-10
    12.3.5 I Edit Descriptors 12-10
    12.3.6 L Edit Descriptor 12-11
    12.3.7 Sign Control Edit Descriptors 12-12
    12.3.8 Slash Edit Descriptor 12-13
    12.3.9 Tabulator Edit Descriptors 12-13

13 PROGRAM UNITS AND SUBPROGRAMS   13-1

13.1 Main Program 13-2

13.2 Modules 13-3

    13.2.1   USE Statement 13-5

    13.2.2   Typical applications 13-7

13.3 Subprograms 13-8

    13.3.1   Module Functions 13-9

        13.3.1.1   Function Definition 13-11

        13.3.1.2   Explicit Function Reference, Invocation 13-14

        13.3.1.3   Operator Functions 13-16

    13.3.2   Module Subroutines 13-16

        13.3.2.1   Subroutine Definition 13-17

        13.3.2.2   Explicit Subroutine Reference, CALL Statement 13-19

        13.3.2.3   Assignment Subroutines 13-20

    13.3.3   External Subprograms 13-20

    13.3.4   Dummy Subprograms 13-21

    13.3.5   Interface Blocks 13-21

    13.3.6   Overloaded Generic Subprogram Names 13-24

    13.3.7   Return from an Invoked Module Subprogram 13-25

13.4 Internal Program Communication 13-27

    13.4.1   Argument Lists 13-27

        13.4.1.1   Dummy Argument List 13-27

        13.4.1.2   Actual Argument List 13-28

    13.4.2   Argument Association 13-29

        13.4.2.1   Data Objects as Dummy Arguments 13-31

        13.4.2.2   Implicit Association of Two Dummy Arguments 13-32

        13.4.2.3   Length of Character Dummy Arguments . . . 13-32

        13.4.2.4   Scalar Arguments 13-33

        13.4.2.5   Dummy (Argument) Arrays 13-33

        13.4.2.6   Dummy (Argument) Pointers 13-35

        13.4.2.7   Restrictions on the Association of Data Entities 13-35

        13.4.2.8   Dummy Subprograms 13-36

    13.4.3   Optional Dummy Arguments 13-36

    13.4.4   Dummy Argument with INTENT Attribute 13-37

14 INTRINSIC SUBPROGRAMS   14-1

14.1 Intrinsic Functions 14-1

    14.1.1 Table of Intrinsic Functions 14-2

14.2 Intrinsic Subroutines 14-6

14.3 Intrinsic Subprogram Reference 14-6

14.4 Intrinsic Subprogram Definitions 14-8

Appendices

A ASCII CHARACTER SET AND COLLATING SEQUENCE   **A-1**

B MODELS FOR NUMBERS   **B-1**
    B.1 Models for Integers B-1
    B.2 Models for Reals B-1
    B.3 Models for Bit Manipulation B-2

C PROGRAM EXAMPLE   **C-1**

D F versus Fortran 90   **D-1**

E F LANGUAGE SYNTAX CHARTS   **E-1**
    E.1 Notation Used in this Syntax E-1
    E.2 F Terms and Concepts E-2
    E.3 Characters, Lexical Tokens, and Source Form E-4
    E.4 Intrinsic and Derived Data Types E-8
    E.5 Data Object Declarations and Specifications E-13
    E.6 Use of Data Objects E-18
    E.7 Expressions and Assignment E-23
    E.8 Execution Control E-28
    E.9 Input/Output Statements E-32
    E.10 Input/Output Editing E-36
    E.11 Program Units E-39
    E.12 Procedures E-42

F INDEX   **F-1**

1 SOURCE FORM

An F program is a collection of *program units*. It consists of one *main program* and any number of *modules*. Each of these program units is a sequence of program lines consisting of F *statements* and/or *comments*. The characters in these program lines form *lexical tokens*. The following rules govern the form of the program lines in a program unit.

A **program line** is a sequence of characters. The character positions of a program line are counted from the left to the right beginning with 1.

With few exceptions, a program line must contain only characters of the F *character set*. The exceptions are *literal constants* and *comments*, which may contain certain other characters (but no control characters) of the *ASCII character set*.

The F **character set**, which is embedded in the ASCII character set, consists of the following characters:

Letters: a b c d e f g h i j k l m n o p q r s t u v w x y z

A B C D E F G H I J K L M N O P Q R S T U V W X Y Z

Digits: 0 1 2 3 4 5 6 7 8 9

Special characters:

			Blank	:	Colon
=	Equals	+	Plus	−	Minus
*	Asterisk	/	Slash	(	Left Parenthesis
)	Right Parenthesis	,	Comma	.	Point
'	Apostrophe	!	Exclamation Point	"	Quote
%	Percent	&	Ampersand	;	Semicolon
<	Less Than	>	Greater Than	?	Question Mark
$	Currency Symbol	_	Underscore		

Each program line has *maximal* 132 characters. A statement may occur anywhere in the program line.

An **initial line** is a program line which is no blank line and has no "!" in its first nonblank character position.

A statement may be **continued** on the next line. The character "&" as the last nonblank character in a line (and not as a part of a comment) indicates that the statement continues at position 1 of the next statement line. A statement must have not more than 39 continuation lines. A program line must not contain an "&" as the first nonblank character.

A keyword, a name, a literal constant, an operator, "=>", "(/", or "/)" are **lexical tokens**. Except a complex literal constant whose real part and imaginary part may be split onto two lines, lexical tokens must not be continued from one line to the other. Compound statement keywords with embedded blanks, such as END IF, cannot be separated and continued.

Spelling of lexical tokens: Except character literal constants and user-defined names, lexical tokens containing characters must be written in lower case. User-defined names may be written in mixed lower and upper case, but all references to a particular name must use the same spelling within the scope of this name. All intrinsic subprogram names and their argument names must be written in lower case.

Blanks must not appear within lexical tokens except within character literal constants. But blanks must be used as separators between keywords, names, and constants and subsequent keywords, names, and constants. Two or more blanks used as separators have the same interpretation as one blank. In the following list of "compound keywords", blanks are optional to separate adjacent keywords:

ELSE IF	END DO	END FILE	END FUNCTION
END IF	END INTERFACE	END MODULE	END PROGRAM
END SELECT	END SUBROUTINE	END TYPE	END WHERE
IN OUT	SELECT CASE		

Because blanks must be used as separators or may be used as meaningful characters in character literal constants, we say that they are "significant". Blanks must not appear in format specifications except on either side of a comma, either side of a parenthesis, after a repeat count, before a field width, or before a tabulator count.

A **comment** serves only documentation purposes and has no effect on the interpretation of the program. A comment may contain any characters of the ASCII character set except ASCII control characters. Lines containing only blanks are comment lines. Comment lines may appear anywhere before the END statement of a program unit or subprogram.

An "!" character not appearing within a character literal constant indicates the beginning of a **comment**. This comment begins at the "!" and ends at the last nonblank character of the line before position 133.

A line with an "!" as the first nonblank character is a **comment line**. A comment line must not be continued.

The following program serves only to demonstrate the spelling of **F** source texts:

```
!00000000000000000000000000000000000000000000000000...111111111111111
!00000000111111111112222222222333333333444444444...112222222222333
!23456789012345678901234567890123456789012345678...890123456789012
program pascal_triangle                          ! initial line
integer, dimension (13) :: basis = 1             ! initial line
integer                 :: null = 23             ! initial line
character (len = 11)    :: format                ! initial line
                                                 ! comment line
write (unit=*, fmt="(tr5,a,tr5,i6/tr22,i6,i5)") & ! initial line
   "Pascal Triangle"              &              ! contin. line
   ! writes the first 3 elements                 comment line
   basis(13), basis(12), basis(13)              ! contin. line
                                                 ! comment line
do i=12,2,-1                                     ! initial line
  do j=i,12                                      ! initial line
    basis(j) = basis(j) + basis(j+1)            ! initial line
  enddo                                          ! initial line
  null = null - 2                                ! initial line
  write (unit=format, fmt="(a3,i2,a6)") &       ! initial line
                     "(tr", null, ",13i5)"      ! contin. line
  write (unit=*, fmt=format) basis(i-1:13)      ! initial line
enddo                                            ! initial line
end program pascal_triangle                      ! initial line
```

This program produces the following output:

```
    Pascal Triangle          1
                          1     1
                       1     2     1
                    1     3     3     1
                 1     4     6     4     1
              1     5    10    10     5     1
           1     6    15    20    15     6     1
        1     7    21    35    35    21     7     1
     1     8    28    56    70    56    28     8     1
  1     9    36    84   126   126    84    36     9     1
1    10    45   120   210   252   210   120    45    10     1
 1    11    55   165   330   462   462   330   165    55    11     1
  1    12    66   220   495   792   924   792   495   220    66    12     1
```

Summary:

Position(s)	Character	Interpretation
1	!	Indicates a comment line
1 – 132	!	Initiates a comment
1 – 132	F character set	F statement
2 – 132	&	Indicates continuation
1 – 132	Only blanks	Comment line

1.1 Classification of F Statements

An F statement is either an *executable statement* or a *nonexecutable statement*.

There are three big classes of F statements: nonexecutable *specification statements*, executable *execution control statements*, and executable *input/output statements*. They are presented in chapter 9, chapter 10, and chapter 11, respectively. And there is a class of *other executable statements* in context with assignments, subprograms, and memory management.

The following statements are **executable statements**. They cause actions of the program:

ALLOCATE	assignment	BACKSPACE	CALL
CASE	CASE DEFAULT	CLOSE	CYCLE
DEALLOCATE	DO	ELSE	ELSE IF
ELSEWHERE	END DO	END FILE	END FUNCTION
END IF	END PROGRAM	END SELECT	END SUBROUTINE
END WHERE	EXIT	IF THEN	INQUIRE
NULLIFY	OPEN	pointer assignment	PRINT
READ	RETURN	REWIND	SELECT CASE
STOP	WHERE	WRITE	

All other F statements are **nonexecutable statements**.

Several executable statements may appear only in combination with certain other statements to form **executable constructs**.

CASE construct:	SELECT CASE, CASE, END SELECT
DO construct:	DO, CYCLE, EXIT, END DO
IF construct:	IF THEN, ELSE IF, ELSE, END IF
WHERE construct:	WHERE, ELSEWHERE, END WHERE

CASE, DO, and IF constructs control the execution of one or more statement blocks. WHERE constructs control the selection of array elements for certain assignment operations. Executable statements which are not used to form constructs are **simple executable statements**.

1.2 Statement Ordering

The following diagram gives an overview of the ordering of statements within program units and subprograms.

A main program is enclosed by PROGRAM and END PROGRAM, a module by MODULE and END MODULE, a function by FUNCTION and END FUNCTION, and a subroutine by SUBROUTINE and END SUBROUTINE.

Note that a user-defined F subprogram must be embedded as a module subprogram into the module subprogram part of a module.

The language elements appearing in a particular box within the following diagram may occur in any order. Note that some statements are allowed only in certain program units, in certain subprograms, or in certain constructs. Horizontal lines separate groups of language elements which must not be interspersed.

PROGRAM	MODULE	FUNCTION / SUBROUTINE
USE statements	USE statements	USE statements
IMPLICIT NONE	IMPLICIT NONE	
	default accessibility: PUBLIC or PRIVATE	
	PUBLIC statements PRIVATE statements	
INTRINSIC stmts.	INTRINSIC stmts.	INTRINSIC statements
		type declarations: dummy arguments interface blocks: dummy arguments
		type declaration: RESULT variable
type declarations	type declarations type definitions interface blocks	other type declarations
executable statements executable constructs		executable statements executable constructs
	CONTAINS statement module subprograms	
END PROGRAM	END MODULE	END FUNCTION / END SUBROUTINE

All type declaration statements and the other specification statements must precede all executable statements. The order of the specification statements within a box may be important if a particular data entity is specified or referenced in more than one specification statement.

Main programs and modules must appear in such an order that the F processor has already processed a particular module before any reference (i. e. USE statement) to that module.

A comment line appearing before a PROGRAM, MODULE, FUNCTION, or SUBROUTINE statement belongs to the main program, module, function, and subroutine, respectively. A comment line subsequent to an END statement is not part of the main program, module, function, and subroutine, respectively, with this END statement.

The END statement is always the last physical line of a program unit or subprogram.

2 TYPE CONCEPT

There are *intrinsic types* and *derived types*. Intrinsic types are provided by the F language, whereas derived types are provided by the programmer. All properties of the intrinsic types are known at any point in an F program.

Each intrinsic type except character type is parameterized. By specifying a **kind type parameter** value for a parameterized intrincic data type, the programmer selects a particular kind of that data type. If the programmer does not explicitly specify a kind type parameter, a default kind type parameter value is assumed, which selects the **default type**, i. e., the default kind of that data type.

For a parameterized intrinsic type, the set of valid values, their internal representation, and their approximation method depend on the value of the selected kind type parameter. Note that the set of supported kind type parameter values is processor-dependent.

In addition to these intrinsic types, the programmer can derive types from intrinsic types and other derived types.

2.1   Intrinsic Types

There are 5 intrinsic types. These are the

numeric types   integer, real, and complex, and the
nonnumeric types   logical and character.

2.1.1   Integer Type

Name: INTEGER.

Sets of values: The set of values of an integer type is a subset of the mathematical integers. Each such set of values has a processor-dependent smallest negative value and a largest positive value.

An F processor must provide at least one internal representation method of integer values. Each of these representation methods is characterized by a processor-dependent kind type parameter value. The programmer may specify (for instance, in addition to the keyword INTEGER) a kind type parameter for an integer data object. If a kind type parameter is not explicitly specified, the default kind type parameter KIND(0)[1] is assumed. This default kind type parameter selects the **default integer type**. Each set of integer values contains the integer zero which is neither positive nor negative. A signed zero and an unsigned zero have the same value.

---

[1] KIND is an intrinsic function. This is a portable way to denote a kind type parameter value. Kind type parameter values even for the default types are processor-dependent (see above).

External representations: An integer value may be represented as an *integer literal constant* with or without a kind type parameter.

Operations: Addition, subtraction, multiplication, division, exponentiation, negation, and identity. These operations are defined for all numeric types.

```
integer, parameter :: small = selected_int_kind(5)
```

The value of **small** is the kind type parameter value of an integer type that has a minimal range from -10^5 to $+10^5$.

```
integer (kind = small) :: x, y      ! is a type declaration
```

2.1.2   Real Type

Names: REAL.

Sets of values: The set of values of a real type is a subset of the mathematical real numbers. Most real numbers have no exact internal representation. Therefore, a real value has a processor-dependent internal representation, which is an approximation to the exact mathematical number.

Such an approximation has two characteristic processor-dependent properties: the (decimal) *precision* and the (decimal) *exponent range*. An F processor must support at least two different approximation methods of internal representation of real values. Therefore, the real type has at least two different sets of values. These two sets of values must be different with regard to precision, and they also may be different with regard to their range.

The approximation methods are characterized by a processor-dependent kind type parameter value. The programmer may specify (for instance, in addition to the keyword REAL) a kind type parameter for a real data object. If a kind type parameter is not explicitly specified, the default kind type parameter KIND(0.0) is assumed. This default kind type parameter selects the **default real type.**

The sets of values contain the real zero. A signed zero and an unsigned zero have the same value.

External representations: A real value may be represented as a *real literal constant* with or without a kind type parameter.

Operations: Addition, subtraction, multiplication, division, exponentiation, negation, and identity. These operations are defined for all numeric types.

```
integer, parameter :: big = selected_real_kind(14, 200)
```

The value of **big** is the kind type parameter value of a real type with a decimal precision of at least 14 decimals and a minimal exponent range from 10^{-99} to 10^{+99}.

```
real (kind = big) :: a, b           ! is a type declaration
```

2.1.3 Complex Type

2

Name: COMPLEX.

Sets of values: The set of values of a complex type is a subset of the mathematical complex numbers. A complex value has an internal representation consisting of two real values, one for the real part and one for the imaginary part.

An F processor must support at least the same approximation methods of internal representation of the real part and imaginary part as in the case of the real type. Both parts of a complex value must be represented according to the same approximation method.

The approximation methods are characterized by a processor-dependent kind type parameter value. The programmer may specify (for instance, in addition to the keyword COMPLEX) a kind type parameter. If a kind type parameter is not explicitly specified, the default kind type parameter KIND(0.0) is assumed. This default kind type parameter selects the **default complex type** and the F processor uses the same approximation method of internal representation of the real and the imaginary part as in the case of the default real type.

The sets of values contain the complex zero, which is neither positive nor negative. A signed zero and an unsigned zero have the same value.

External representations: A complex value may be represented as a *complex literal constant* with or without a kind type parameter.

Operations: Addition, subtraction, multiplication, division, exponentiation, negation, and identity. These operations are defined for all numeric types.

2.1.4 Logical Type

Name: LOGICAL.

Sets of values: The set of values of a logical type contains only two values with the interpretation *true* and *false*, respectively.

An F processor must support at least one internal representation method of logical values. Each of these representation methods is characterized by a processor-dependent kind type parameter value. The programmer may specify (for instance, in addition to the keyword LOGICAL) a kind type parameter. If a kind type parameter is not explicitly specified, the default kind type parameter KIND(.FALSE.) is assumed. This default kind type parameter selects the **default logical type**.

External representations: A logical value may be represented as a *logical literal constant* with or without a kind type parameter.

Operations: Negation, conjunction, inclusive disjunction, logical equivalence, and logical non-equivalence.

2.1.5   Character Type

Name: CHARACTER.

Set of values: The set of values of the character type contains *character strings*. A **character string** is a sequence of characters. Each character in a string has a position within this string. The positions are numbered from the left to the right beginning with 1, 2, 3, ... The number of the last character position of a string is equal to the **length** of the string. The length may be zero.

An F processor provides one internal representation method of character values: the ASCII code. A character string may contain any *representable character* of the ASCII character set. A kind type parameter cannot be written to select another character set.

External representations: A character value (that is, a character string) may be represented as a *character literal constant*.

Operation: Concatenation.

2.2   Derived Types

Additional data types may be defined by the programmer. For example, a new data type may be derived from intrinsic types; and such a derived type may be used to define another derived type. A derived type has at least one *type component*. Each component of a derived type specifies an intrinsic type or a derived type.

Name: The name of a derived type is defined in the TYPE definition statement.

Set of values: The set of values of a derived type is a combination of the sets of values of the components of this data type. Ultimately, a value of a derived type is a collection of values of intrinsic types.

External representation: The external representation of the values of a derived type is given by a method for the construction of these values; it is called a *structure constructor*.

Operations: There are no intrinsic operations (except the assignment operation). Additional operations may be defined by the programmer with help of *operator functions* and *operator interface blocks*.

2.2.1   Derived Type Definition

A derived type definition is used to define the name of a new derived type and to define the names, attributes, and types of the components of this new derived type. Such a derived type definition begins with a TYPE definition statement, it ends with an END TYPE statement, and contains at least one type component definition.

2

TYPE, $\left\{ \begin{array}{l} \textbf{PUBLIC} \\ \textbf{PRIVATE} \end{array} \right\}$ **:: type_name**
  [**PRIVATE**]
  component_definition
  [**component_definition**]
  ⋮
END TYPE type_name

If PUBLIC is specified within the TYPE definition statement, a PRIVATE statement (without a list) may appear after the TYPE definition statement but before the first component definition. In this case, the component names and thus the internal structure of the derived type are available only within the module containing the derived type definition.

TYPE definition statement

A derived type definition begins with a TYPE definition statement.

TYPE, PRIVATE :: type_name

TYPE, PUBLIC :: type_name

The **type_name** must not be the same as the name of an intrinsic type or the name of another derived type being available in the scope of the defined derived type. But it may be same as the name of one of its pointer components, if any. PRIVATE means that the derived type is available only within the scoping unit of the module containing this derived type definition; there are no means to make such a derived type accessible outside the module. And PUBLIC means that the derived type may become accessible outside the module. A public type also is described as a "visible" type.

Equality of derived data types

Two data entities in a scoping unit have the same derived type if they are declared with regard to the same derived type definition. Two data entities in different scoping units have the same data type if they are declared with regard to the same derived type definition, which may be accessed from a module or from the host.

Two data entities in different scoping units are not of equal derive type if at least one of these entities is private or has a private component.

2.2.1.1 Type Component Definition

The definition of a type component has a form similar to a type declaration statement. A component may be defined to have the DIMENSION attribute and/or the POINTER attribute.

type [, **attribute** [, **attribute**]] :: **type_component** [, **type_component**]...

type is a *type specification* of a derived type or of an intrinsic type with or without a kind type parameter specification. And **type_component** is a the name of a type component.

A character length specification which is included in the type specification of a character component must be a constant specification expression. As an **attribute**, only **POINTER** and/or **DIMENSION (dim** [**, dim**]... **)** may be specified.

If the **type** is a derived type and POINTER is not specified, this **type** must be defined earlier in the scoping unit or must be accessed by USE association or host association. If the **type** is a derived type and POINTER is specified, this **type** may be any available derived type or even the derived type being defined.

```
type, public :: date
  integer          :: day
  character (len=3) :: month
  integer          :: year
end type date
```

The definition of the derived type **date** contains three type components. In a scoping unit where this derived type is available, data objects of type **date** may be declared:

```
type (date) :: birthday, holiday, tax
```

An array object may be declared to be of a derived type:

```
type (date), dimension (25, 13) :: pupil
```

The variable **pupil** is a 2-dimensional array of type **date**. The value of any element of array **pupil** represents a date consisting of three values, as is specified in the derived type definition.

Suppose, array **pupil** contains the birthdays of pupils. Then the complete date for the 3rd pupil in classroom no. 12 or a part of that date may be referenced:

```
print *, pupil(3, 12)      ! prints the complete date
print *, pupil(3, 12)%year ! prints only the year of the birthday
```

Array components

A type component with DIMENSION attribute is an **array component**. If the array component is not a *pointer component*, the array specification has the same form as for an *explicit-shape array* and the array bounds must be constant *specification expressions*. If the array component also has the POINTER attribute, the array specification has the same form as for an *array pointer*.

```
type, public :: experiment
  integer               :: number
  type (date)           :: day
  real, dimension (100) :: sensor1, sensor2
  real, dimension (24)  :: hour
end type experiment
```

Where this derived type is available, variables for the storage of measurement results may be declared:

```
type (experiment) :: temperature, density, height
```

Pointer components

A type component with POINTER attribute is a **pointer component**.

```
type, public :: bibliography
  integer                                     :: volume, year, pages
  character (len=72)                          :: titel
  character (len=1), dimension (:), pointer :: abstract
end type bibliography
```

In a scoping unit where this derived type definition of **bibliography** is available, objects of this type may be declared. Such an object has four components with known memory requirements. These are the default integer components **volume**, **year**, and **pages** and the character component **titel**. And there is an additional component **abstract**, which is a pointer that may become pointer associated with a 1-dimensional character array.

A pointer component in a derived type definition is allowed to have the same type as the type being defined:

```
type, public :: link
  integer             :: position
  type (link), pointer :: left, right
end type link
```

Where this derived type is available, objects of type **link** may be declared. These objects may be manipulated as elements of a linked list.

2.2.1.2   Private and Public Derived Types and Components

A derived type (definition) or a type component (definition) is **private** if it is accessible only in the module containing the derived type definition. Outside the module, a private type component and a private derived type are not accessible and there are no means to make them accessible.

A derived type (definition) or a type component (definition) is **public** if it may also be made accessible outside the module containing the derived type definition. Public entities also are described as "visible" entities. Such entities of a module may be made accessible outside the module by use of a USE statement.

A derived type is private if the PRIVATE attribute is specified in its TYPE definition statement

A derived type is public if the PUBLIC attribute is specified in its TYPE definition statement.

If a data type is private, the following characteristics and concepts of this type are available only in the module containing the derived type definition: the name of the type, the names of its components, objects and structure constructors of this type, and subprograms with dummy arguments of this type and function results of this derived type.

A component of a derived type is private if it is a component of a private type or if the derived type definition contains a PRIVATE statement or if the type of the component is another private type. If at least one component of a derived type is private, all components must be private.

```
type, public :: point
  private
  real :: x, y
end type point
```

The type **point** is available within the module and by a USE statement outside the module. The components **x** and **y** are only available within the module; the inner structure of the type **point** cannot be made accessible outside the module.

```
type, private :: note
  integer           :: number, weight
  logical           :: ok
  character (len=72) :: text
end type note
```

This type **note** is not visible. It cannot be made accessible outside the module.

2.2.2   Structure Objects

2

Structure objects are *scalar* entities of a derived type such as *structure variables* and *structure constructors*.

The name of a *type component* is needed outside the derived type definition only when a single component of a structure object is referenced by qualifying the name of the parent structure by the name of the corresponding type component.

Structure variables

A structure variable must be declared by a TYPE declaration statement. Note that the type definition of this derived type must be *available* in the same scoping unit before this TYPE declaration statement. "Available" means that the derived type definition must either appear before the declaration or must be accessible by *USE association* or *host association*.

Structure constructor

In a scoping unit where the derived type definition and the inner structure of the derived type are available, values of this derived type may be constructed. A **structure constructor** supplies a series of values:

type_name (expression [, expression]...)

The list of expressions enclosed in parentheses must supply a value for each single component of the derived type **type_name**. And these values must be suitable (in number and order) for the components of this derived type. If the result of such an expression does not agree in type, (if applicable) in kind type parameter, or (if applicable) in character length with those of the corresponding type component, the result is converted according to the rules for intrinsic assignment statements. For a nonpointer component, the shape of the result of the expression also must conform with the shape of the component.

A structure constructor may appear only after the derived type definition.

```
type, public :: string
  integer              :: length
  character (len = max) :: line
end type string

character (len=25) :: text
character (len=8)  :: mark
type (string)      :: colour
```

Where these specifications are available, the following statements may be executed:

```
read *, text, mark
colour = string(len(text) + len(mark), text // mark)
```

An expression is given for each component of the type **string**. After evaluation of these expressions, the constructed value is assigned to the variable **colour**. This variable on the left-hand side of the assignment statement has the same type as the structure constructor on the right-hand side.

If the derived type has a pointer component, the **expression** corresponding to this pointer component must evaluate to a result which would be legal on the right-hand side of a *pointer assignment statement*.

```
type (bibliography) :: book
character (len=1), dimension (1000), target :: source
 ⋮
book = bibliography(1, 1997, 350, "The F Language Guide", source)
```

In this example, **source** is the target object for the corresponding component **abstract**; compare an earlier example in the section on "Pointer components".

Constant structure constructors

If all expressions in a structure constructor are constant expressions, it is a **constant structure constructor** or more precisely a *derived type constant expression* or simply a *structure constant*.

```
type, public :: date
  integer        :: day
  character (len=3) :: month
  integer        :: year
end type date
```

Where this typdefinition is available, for instance by USE association or host association, a structure constant of type **date** may be used as in the following statements:

```
type (date) :: birthday
 ⋮
birthday = date(11, "OCT", 1996)
```

Within the structure constructor, a constant is given for each component. The structure constant is assigned to the variable **birthday**, which also is of type **date**.

3 LEXICAL TOKENS

An F statement consists of **lexical tokens**. These are keywords, names, operators, literal constants (except complex literal constants), delimiter pairs as (...), /.../, " ...", ..., and (/.../), and finally, =, =>, &, :, ::, ;, and %.

3.1 Scoping Units

A program unit consists of one or more nonoverlapping *scoping units*. A **scoping unit** consists of all program lines of a derived type definition, of an interface block without any of its contained interface blocks, or of a subprogram without any of its contained interface blocks, or of a program unit without any of its contained derived type definitions, interface blocks, and subprograms.

The **scope** of a lexical token is that part of a program where the interpretation of the token is unambiguous. If the scope of a lexical token is the program, the token has a **global scope**. And if the scope is a scoping unit, the lexical token has a **local scope**. There are lexical tokens with a scope that is only a part of a statement.

3.2 Keywords

Keywords are the given unmodifiable parts of the statements, called "statement keywords", and the names of the dummy arguments, called "argument keywords". Most of the statement keywords are *reserved words* in F (see below "Names").

3.3 Names

Names identify data objects, subprograms, data types, and so on. Normally, names are invented and inserted by the programmer. The scope of a name contains all statements where the name is known and usable. Names with global scope are *global names*. And names with local scope are *local names*. There are *special names* with a scope consisting only of a part of a statement. Normally, a name is unambiguous in its scope (exception: generic names).

Regardless of case, a programmer invented name must not agree with a *reserved word*. The list of reserved words of the F language contains the logical operators (for instance, **eqv**), die logical literal constants (for instance, **true**), the names of the intrinsic subrpograms (for instance, **exp**), the names of certain standard subprograms not being supported by the F language (for instance, **dble**), the historical names of the relational operators not being supported by the F language (for instance, **eq**), certain names and keywords being reserved for Fortran 95 (for instance, **null**, **forall**), and certain statement keywords being contained in Fortran 90 but not in the F language (for instance, **continue**). In addition, the list of reserved words contains all statement keywords, including the attribute specifiers in type declaration statements, excluding the specifiers in input/output statements and the keyword **stat**.

The complete **list of reserved words** contains:

abs	achar	acos	adjustl
adjustr	aimag	aint	all
allocatable	allocate	allocated	and
anint	any	asin	assignment
associated	atan	atan2	backspace
bit_size	btest	call	case
ceiling	char	character	close
cmplx	complex	conjg	contains
continue	cos	cosh	count
cpu_time	cshift	cycle	date_and_time
dble	deallocate	default	digits
dim	dimension	do	dot_product
dprod	elemental	else	elseif
elsewhere	end	enddo	endfile
endforall	endfunction	endif	endinterface
endmodule	endprogram	endselect	endsubroutine
endtype	endwhere	eoshift	epsilon
eq	eqv	exit	exp
exponent	false	floor	forall
fraction	function	ge	go
goto	gt	huge	iachar
iand	ibclr	ibits	ibset
ichar	ieor	if	implicit
in	index	inout	inquire
int	integer	intent	interface
intrinsic	ior	ishft	ishftc
kind	lbound	le	len

len_trim	lge	lgt	lle
llt	log	log10	logical
lt	matmul	max	maxexponent
maxloc	maxval	merge	min
minexponent	minloc	minval	mod
module	modulo	mvbits	ne
nearest	neqv	nint	none
not	null	nullify	only
open	operator	optional	or
out	pack	parameter	pointer
precision	present	print	private
procedure	product	program	public
pure	radix	random_number	random_seed
range	read	real	recursive
repeat	reshape	result	return
rewind	rrspacing	save	scale
scan	select	selectcase	selected_int_kind
selected_real_kind	set_exponent	shape	sign
sin	sinh	size	spacing
spread	sqrt	stop	subroutine
sum	system_clock	tan	tanh
target	then	tiny	to
transfer	transpose	trim	true
type	ubound	unpack	use
verify	where	write	

A **name** consists of one through 31 letters, digits, and underscores "_". The first character must be a letter and the last one must not be an "_". Though case is not significant for the spelling of user-defined names, only one written form of a name is allowed within a scoping unit. The names of intrinsic subprograms and their dummy arguments must be written in lower case.

salary	peter	x12	e605	FReferenceManualPages
UPPER_CASE		← single underscore character		
b_o_l_d__face		← consecutive underscore characters		
DINA4	DinA4	← must not appear within the same scoping unit		

Global names identify the following **global entities**: main programs and modules.

Local names identify **local entities**, which are classified as follows:

1. Named variables (except variables with *special names*, see below), named constants, named constructs, module subprograms, dummy subprograms, intrinsic subprograms, interface blocks with generic names, derived types,

2. Type components, and

3. Argument keywords

Note that the names of type components form a separate class for each derived data type and that the argument keywords form a separate class for each subprogram.

Normally, a name that identifies a global entity in a scoping unit must not be used to identify a local entity of class 1. (see above) in that scoping unit. A local name may identify in the *same* scoping unit another local entity of *another* class. A local name may identify in *other* scoping units another local or global entity.

Special names: A **special name** has a scope that consists only of a part of a statement. The name of a *DO variable* of an implied-DO in an array constructor has a scope that is the implied-DO list. Such names may be reused in the scoping unit but only as DO variables in array constructors.

3.4   Operators and Assignment Symbol

Intrinsic operators are global entities. *Defined operators* are local entities. Two different operations in a scoping unit may use the same defined operator; this is called "overloading".

The assignment symbol is a global entity. Within a scoping unit, the assignment symbol may identify additional *defined* assignment operations or replace the intrinsic derived type assignment operation.

3.5   Literal Constants

There are five forms of *intrinsic* literal constants: integer, real, complex, logical, and character literal constants. For each kind of an intrinsic data type, an F processor supports a literal constant which has its own external representation and differs from other literal constants with regard to internal representation and/or internal approximation method.

The type of a literal constant is not specified explicitly but follows from the form of the constant. Integer, real, and complex constants are **numeric constants**.

3.5.1   Integer Literal Constants

An integer literal constant is a string of digits. It consists of one or more digits, an optional sign, and an optional trailing underscore followed by a kind type parameter. Positive integer literal constants *may* be written with a leading sign "+". Negative integer literal constants *must* be written with a leading sign "−".

[±] **d** [**d**]... [_ **kind**]

Each **d** is a digit. **kind** is a processor-dependent kind type parameter value, which must be written as a scalar nonnegative integer named constant.

Such an integer constant is interpreted as a decimal value. If there are particular requirements concerning the range of an integer literal constant, a kind type parameter may be written. The value of the kind type parameter must be supported by the F processor. If a kind type parameter value is not written, the constant is of type default integer.

```
1    531    +76    9876    -239657    0       ←── type default integer
100_bi       1024_dec        739840_b4    ←── with kind type parameter
```

3.5.2   Real Literal Constants

A real literal constant is a string of digits containing a point. Positive real literal constants and positive exponents *may* be written with a leading sign "+". Negative real literal constants and negative exponents *must* be written with a leading sign "−". A real literal constant may be written with a trailing underscore followed by a kind type parameter.

[±] **dec** [_ **kind**]

[±] **dec** E [±] **exponent** [_ **kind**]

dec is a decimal number of the form **n.n**, where **n** is an unsigned integer literal constant. The **exponent** is an integer literal constant; base is 10. **kind** is a processor-dependent kind type parameter value, which must be written as a scalar nonnegative integer named constant.

If there are particular requirements concerning the decimal precision and/or the decimal exponent range of the internal representation, a kind type parameter may be written. The value of the kind type parameter must be supported by the F processor. If a kind type parameter value is not written, the constant is of type default real.

Constants of type default real:

12.0   -12.34   +0.34

12.0e-2   value: $12.0 * 10^{-2} = 0.12$
-12.34e+3   value: $-12.34 * 10^3 = -12340.0$
+0.34e4   value: $0.34 * 10^4 = 3400.0$

Real constants with processor-dependent kind type parameter:

-23.4_d4   3.0e-5_d10   3.75e7_b7

3.5.3   Complex Literal Constants

A complex literal constant is written as a pair of real literal constants, which are separated by a comma and enclosed in parentheses.

(real_part, imaginary_part)

The **real_part** and the **imaginary_part** are real literal constants.

The parentheses are parts of the constant. If a kind type parameter is written within one part, it must also be written for the other part with the same named constant.

Complex constants of type default complex:

(7.0, 3.14)   value: $7.0 + 3.14\,i$
(-6.2e-3, 9.0)   value: $-0.0062 + 9.0\,i$

Complex constants with processor-dependent kind type parameter:

(-6.2e-3_d8, 9.0_d8)   value: $-0.0062 + 9.0\,i$

3.5.4   Logical Literal Constants

There are exactly two logical literal constants. One represents the value *true* and the other represents the value *false*; their source text form is .true. and .false., respectively. A logical literal constant may be written with a trailing underscore followed by a kind type parameter. The points are parts of the constants.

.TRUE.[_kind]   for the value *true*.

.FALSE.[_kind]   for the value *false*.

kind is a processor-dependent kind type parameter value, which must be written as a nonnegative integer named constant.

If there are particular requirements concerning the internal representation, a kind type parameter may be written. The value of the kind type parameter must be supported by the F processor. If a kind type parameter value is not written, the constant is of type default logical.

```
.true.          ← default logical constant
.false.         ← default logical constant
.false._b4      ← with processor-dependent kind type parameter
```

3.5.5   Character Literal Constants

A character literal constant is a character string enclosed either in quotation marks. The string may be empty, that is, the string may consist of no character at all. Blank characters within character constants are significant; they are parts of the character value and contribute to the length of the character constant. The delimiting quotation marks are not parts of the value of the character literal constant.

`"[c]..."`

Each **c** is a character of the F character set or another *representable character* (see below) of the ASCII character set.

A quotation mark " " "appearing within the string of characters is represented by two consecutive quotation marks without any intervening blanks if the string is delimited by quotation marks. The included pair of quotation marks is counted as one character.

The number of characters between the delimiters is the **length** of the character constant. The length is fixed and may be zero. A zero-length character literal constant is represented by two consecutive quotation marks without any intervening blanks.

A **representable character** is any character of the ASCII character set except a control character.

```
"American National Standard"  -   length 26
"Programming Language F"       -   length 22
""""                           -   length 1
"3.1415"                       -   length 6
"2,5"" height"                 -   length 11
""                             -   length 0
```

4 DATA OBJECTS

Data entities (or simply **data**) are *constants, variables, results of the evaluation of expressions*, and *function results*. **Data objects** (or simply **objects**) are *constants* and *variables*.

Each data object has a data type. The data types of a literal constant and of a *structure constructor* are implicitly given by their respective forms. The types of a named constant and of a named variable must be specified explicitly.

A constant has *always* a value. This value may be referenced (that means, used), but a constant can never be redefined during the execution of the program. A variable *may* have a value, may have no value, or may have no value at times. Variables may be defined or redefined during the execution of the program.

The name of an object may be used to specify the type and/or additional *attributes* of the object in a type declaration statement.

A scalar derived type object is a *structure object* (or simply a *structure*). A structure has components. These structure components are nonverlapping subobjects of the structure object.

Each object has a *rank*; that is, it is either a *scalar* or an *array*. An object is a *scalar* if it is not an array. A structure is a scalar even if it has an array component.

An *array* is a collection of scalar objects, which have the same type, (if applicable) the same kind type parameter, and (if applicable) the same character length. These *array elements* are ordered (from the point of view of the programmer) such that they form a vector, a matrix, or a cube, and so on.

An *array section* is a subset of the elements of an array. It has (nearly) all properties of an array object, but it has no name. Array sections of a particular array may overlap.

The terms "scalar" and "array" also are used to characterize the *rank* or *shape* of a data entity. For example, the result of the evaluation of an expression or a function result may be scalar or array-valued.

There is a method for the construction of 1-dimensional array values; it is called an *array constructor*.

A *pointer* is a variable which has the POINTER attribute. A pointer may not be referenced or defined until it becomes associated with a *target*.

4.1 Constants

A **constant** has a data type, (if applicable) a kind type parameter, (if applicable) a character length, and a value. If a constant has the PARAMETER attribute, it also has a name. A constant having a name is a **named constant**. *Literal constants* and *constant structure constructors* are constants having no names.

There are five *intrinsic* types of constants: integer, real, complex, logical, and character constants. The data type of a literal constant need not be and cannot be specified explicitly, the type is given by the form of the constant. The data type of a named constant must be specified explicitly.

Integer, real, and complex constants are called **numeric constants**.

Subobjects of constants

A subobject of a constant has the same form as a subobject of a variable. Subobjects of constants are: an array element of a named constant, an array section of a named constant, a structure component of a named constant, a character substring of a named character constant, and a character substring array section of a named character constant.

A subobject of a constant seems to have similar properties as a variable. It may depend on the values of variables, for example, if the subobject designator references a variable in a subscript expression. Or two identical subobject designators of a constant in a scoping unit may denote different values. Or the actual part of the parent constant which is denoted by a subobject designator may not be known until the execution of the program.

Because subobjects of constants are also constants, they must not be redefined.

```
character (len=3) :: command1, command2
character (len=3), dimension (6), parameter :: &
        nix = (/"ls ", "c  ", "mcd", "rm ", "cat", "man"/)
integer :: i
i = 1
command1 = nix(i)(1:1)
  ⋮
i = 6
command2 = nix(i)(1:1)
```

The first `nix(i)(1:1)` designates the value "l". And the second appearance of `nix(i)(1:1)` designates the value "m".

4.2 Variables

A **variable** is

- A named scalar variable;

- A named array variable; or

- An unnamed subobject; that is,
  - An array element (= scalar object);
  - An array section (= array object);
  - A structure component (= scalar or array object); or
  - A substring of a character variable (= scalar object).

A subobject is designated by first writing the parent object, of which it is a part, and then writing additional details, which "qualify" the parent object until the subobject is unambiguously specified. These qualifiers are different for array elements, array sections, etc.

A variable has a type, (if applicable) a kind type parameter, (if applicable) a character length, and, sometimes but not always, a value. There are five *intrinsic* types of variables: integer, real, complex, logical, and character variables. In a scoping unit where a derived type definition is available, named variables of this derived type may be declared. The type of a named variable must be specified explicitly. The integer, real, and complex variables are called **numeric variables**.

A variable may be *defined* or *redefined* during the execution of the program; that is, it may be supplied with a *valid* value or with a new valid value. A variable which has a valid value has the definition status "defined". A variable which does not have a valid value has the definition status "undefined".

A variable appearing in an executable statement is either interpreted as the address of the variable (for example on the left-hand side of a numeric assignment statement) or as the value of the variable (for example on the right-hand side of a numeric assignment statement). The form of the variable is independent of its usage as an address or as a value.

Named variables

The type of a named variable must be specified in a type declaration statement. Each declared object without the PARAMETER attribute is a variable.

Unnamed variables, subobjects

The type of an unnamed variable depends on the type of the parent object of which it is a subobject. A type declaration is neither needed nor possible.

Array element: an array element is of the same type as its parent array.

Array section: an array section is of the same type as its parent array.

Structure component: a structure component is of the type that is specified for its corresponding type component. The type of the type component is explicitly specified in the derived type definition.

Character substring: a character substring is always of type character.

4.3   Scalars

An object which is no array is a scalar. Its value is a single value in the set of values which characterizes the type of the object. Scalars have *rank* zero.

4.3.1   Character Substrings

Parts of a scalar character object may be referenced. A **character substring** (or simply **substring**) is a contiguous portion of its parent string.

The characters of the scalar parent string are counted from the left to the right, beginning with character position 1. For a reference to a substring, its parent string, the starting position of the substring, and the ending position of the substring within the parent string must be specified.

parent_string ([starting_position] : [ending_position])

The **parent_string** must be a scalar named character variable, a character array element, or a scalar structure component of type character. The **substring** expressions **starting_position** and **ending_position** must be scalar expressions. If the **starting_position** is not specified explicitly, the default value 1 is assumed. If the **ending_position** is not specified explicitly, its default value is equal to the length of the parent string (this corresponds to the last position of the parent string).

The results of the substring expressions must be in the range from 1 up to and including the length of the parent string. The **length** of a character substring is the number of characters in the substring, which may be calculated as

$$\text{MAX}(\textbf{ending\_position} - \textbf{starting\_position} + 1, 0).$$

For a *nonzero*-length substring, (that is, if the value of the **ending_position** is not less than the value of the **starting_position**), following inequality holds:

$1 \leq$ **starting_position** $\leq$ **ending_position** $\leq$ length of the **parent_string**.

```
character (len=12), save :: z = "KINDERGARTEN"
```
Suppose, the parent string is the variable **z**, then is

Character string	Value
z (8:10)	ART
z (:4)	KIND
z (:)	KINDERGARTEN

The reference to **z(:)** is the same as the reference to **z**.

```
character (len=6), dimension (3), save :: &
         z_ar = (/"Lunedi", "Sabato", "Giorno"/)
```
Suppose, the parent strings are the array elements **z_ar(1)**, **z_ar(2)**, and **z_ar(3)**, then is

Character string	Value
z_ar (1) (:)	Lunedi
z_ar (2) (5:)	to
z_ar (3) (3:4)	or

The reference to **z_ar(1)(:)** is the same as the reference to **z_ar(1)**.

4.4 Arrays

An **array** is a collection of scalar data, which are rectangularly ordered as a vector, a matrix, or a cube, etc. F supports 1-dimensional, 2-dimensional, 3-dimensional, and multi-dimensional arrays up to seven dimensions. An array has a type, (if applicable) a kind type parameter, (if applicable) a character length, certain characteristics concerning its shape, as there are dimensionality and size, possibly a name, and possibly a value.

Corresponding to the five intrinsic types, there are integer, real, complex, logical, and character arrays. In a scoping unit where a derived type definition is available, arrays of this derived type may be declared. Integer, real, and complex arrays are called **numeric arrays**.

The type of a named array must be specified in a type declaration statement.

An **array section** is a subset of the scalar data (that is, the elements) of its parent array. It is an array object which is unnamed. An array section may be referenced by the qualified name of its parent array. For example, such an array section may be designated by the name of its parent array followed by an *array section subscript list* enclosed in parentheses.

An **array element** is the smallest subobject of an array. It is a scalar and has no name. An array element may be referenced by the qualified name of its parent array. For example, such an array element may be designated by the name of its parent array followed by a *subscript list* enclosed in parentheses.

All elements of an array are of the same type, have the same kind type parameter (if applicable), and have the same character length (if applicable). Type, kind type parameter, and length are those of the parent array. An array element may have a value.

Named arrays must be declared by an *array declaration* specifying the characteristics of the array. The specification of the rank and (if applicable) the shape of an array is called an *array specification*.

An array declaration may contain the specification of *array bounds*. Then a *lower bound* and an *upper bound* may be specified for each dimension.

Array declarations and *array specifications* are presented in detail in chapter 6.

```
integer, dimension (5, -100:0) :: l, m
complex, dimension (5, -100:0) :: n
```

There are also unnamed arrays, for example, the value of an array-valued expression, an array section, or a structure component that is an array.

F supports different kinds of arrays:

Allocatable array: an array with ALLOCATABLE attribute. Its array bounds are not specified in the array declaration but in an ALLOCATE statement. When this ALLOCATE statement is executed, the array will be allocated, that is to say, created. An existing allocatable array may be deallocated.

Array pointer: an array with POINTER attribute. Its array bounds are not specified in the array declaration but, for example, in an ALLOCATE statement for pointer allocation.

Assumed-shape array: a dummy argument array which receives its shape from its associated actual argument array.

Automatic array: an array in a subprogram that is no dummy argument array and which has a shape that depends on (at least) one nonconstant specification expression.

Dummy argument array: an array which is a dummy argument in a subprogram.

Explicit-shape array: an array whose upper array bounds are specification expressions. The shape and the length of such an array are given.

Variable array: an array which has a shape that depends on at least one *non*constant specification expression.

The **size of an array** is the total number of its array elements. The size is equal to the product of the sizes of the dimensions of the array. The size of a dimension is called the **extent** of the array in that dimension. The extent of an array in a dimension is normally calculated as (**upper_bound** − **lower_bound** + 1) (see chapter 6). An array may have size zero.

```
real, dimension (15) :: vector            ! 1-dim., 15 elements
logical, dimension (-5:5, 1976:1985) :: tab ! 2-dim., 11*10 elem.
integer, dimension (-30:40, 1981:1986, 5) :: temp
                                          ! 3-dim., 71*6*5 elem.
```

The number of dimensions of an array is called the **rank** of the array. The **shape** of an array is given by its rank and by the extents of the array in all dimensions. The shape may be described as a 1-dimensional array whose array element values are equal to the extents in the corresponding dimensions. Note that the shape of an array does not say anything about the precise array bounds.

Once an array is declared, its rank remains fixed during the execution of the program. But for a dummy argument array, an automatic array, an array pointer, and an allocatable array, the extents of the array in each dimension and thus its size may vary.

4.4.1 Inner Structure of Arrays

An array is a data object. It consists of a set of scalar array elements, which are ordered rectangularly. This ordering of the array elements as a vector, a square, or a cube, etc. happens only in the imagination of the programmer. In addition to this application-oriented one- or multi-dimensional imaginary structure, there is an *internal* (1-dimensional) imaginary sequence of the array elements. This internal order is called the *array element order*.

The array elements are internally ordered *column-wise*: if a multi-dimensional array is referenced in array element order, the first subscript changes most rapidly and the last subscript changes most slowly.

```
real, dimension (4, 3) :: y  ! 2-dimensional array with 12 elements
```

	Column 1	Column 2	Column 3
Row 1	y (1,1)	y (1,2)	y (1,3)
Row 2	y (2,1)	y (2,2)	y (2,3)
Row 3	y (3,1)	y (3,2)	y (3,3)
Row 4	y (4,1)	y (4,2)	y (4,3)

Array element	y (1,1)	y (2,1)	y (3,1)	y (4,1)	y (1,2)	y (2,2)	⋯
Position	1	2	3	4	5	6	⋯

Array element	y (3,2)	y (4,2)	y (1,3)	y (2,3)	y (3,3)	y (4,3)
Position	7	8	9	10	11	12

```
real, dimension (3, 3, 2) :: z  ! 3-dimensional array with 18 elem.
```

	Col. 1	Col. 2	Col. 3	Col. 1	Col. 2	Col. 3
Row 1	z (1,1,1)	z (1,2,1)	z (1,3,1)	z (1,1,2)	z (1,2,2)	z (1,3,2)
Row 2	z (2,1,1)	z (2,2,1)	z (2,3,1)	z (2,1,2)	z (2,2,2)	z (2,3,2)
Row 3	z (3,1,1)	z (3,2,1)	z (3,3,1)	z (3,1,2)	z (3,2,2)	z (3,3,2)
		Plane 1			Plane 2	

Element	z (1,1,1)	z (2,1,1)	z (3,1,1)	z (1,2,1)	z (2,2,1)	z (3,2,1)	⋯
Position	1	2	3	4	5	6	⋯

Element	z (1,3,1)	z (2,3,1)	z (3,3,1)	z (1,1,2)	z (2,1,2)	z (3,1,2)	⋯
Position	7	8	9	10	11	12	⋯

Element	z (1,2,2)	z (2,2,2)	z (3,2,2)	z (1,3,2)	z (2,3,2)	z (3,3,2)
Position	13	14	15	16	17	18

Array element order

The position of a single array element $(s_1, s_2, \ldots, s_n)$ within the internal (imaginary) sequence of elements is given by the following formula:

$$1 + (s_1 - j_1) + \sum_{m=1}^{n-1} \left((s_{m+1} - j_{m+1}) * \prod_{i=1}^{m} d_i \right)$$

Where n is the rank of the array,

s_i is the integer result of the i-th subscript expression,

j_i is the lower array bound in the i-th dimension,

k_i is the upper array bound in the i-th dimension, and

$d_i = max(k_i - j_i + 1, 0)$ is the extent in the i-th dimension.

If the array has *not* size zero, then $j_i \leq s_i \leq k_i$ for $i = 1, 2, ..., n$.

4

```
real, dimension (0:15) :: r
⋮
r(5) = 94.0
```

The array element `r(5)` is at position 6 in the 1-dimensional array `r`, because $(1 + (5 - 0)) = 6$.

```
character (len=8), dimension (0:2, -2:3) :: s
⋮
s(0, 3) = "examples"
```

Array element `s(0, 3)` is at position 16 in the 2-dimensional array `s`, because $(1 + (0 - 0) + (3 - (-2)) * (2 - (-0) + 1) = 16$.

If the position of an array element and its character length are given, the first character position of an array element relative to the first character position of the array can be calculated as

starting_position $= 1 +$ (array_element_position $- 1$) $*$ character_length.

The array element `s(0, 3)` starts at character position 121, because $1 + (16 - 1) * 8 = 121$.

4.5 Structure Components

A **structure component** is a subobject of a derived type object. It is either one of the components of a (scalar) structure object or it is an array whose elements are themselves components of the corresponding elements of a derived type array.

A structure component does not have a name. It may be referenced by specifying the qualified name of its parent object. Where the name of the type component of the derived type is available, a structure component is identified at least by the designator of the parent object, of which it is a part, and the name of the corresponding type component.

part₁ [%partᵢ]... %partₙ

part₁ is the designator of the parent object of derived type. It is the name of a structure object or of an array object, an array element designator, or an array section designator.

part$_i$ is of derived type. It may be one of the following forms:

- A **component_name** of the preceding **part**$_{i-1}$. This component may have the DIMENSION attribute.
- An array element designator **array (subscript_expr [, subscript_expr]...)**, where **array** is the name of a component (with DIMENSION attribute) of the preceding **part**$_{i-1}$.
- An array section designator **array (section_subscript [, section_subscript]...)**, where **array** is the name of a component (with DIMENSION attribute) of the preceding **part**$_{i-1}$.

And **part**$_n$ is the name of a component of the preceding **part**$_{n-1}$. This component may have the DIMENSION attribute. Not more than one **part**$_k$ may designate a whole array or an array section.

If **part**$_k$ designates a whole array or an array section, the names **component_name** and **array**, appearing in **part**$_{k+1}$ up to **part**$_{n-1}$, and the name **part**$_n$ must not have the POINTER attribute.

Structure components which are array objects, and array sections of structure components also are discussed in chapter 6.

The type of a structure component is given by the type of the component **part**$_n$. If the structure component has a nondefault intrinsic type or the character type, the kind type parameter and the character length, respectively, must be explicitly specified for the type component **part**$_n$ in the derived type definition for **part**$_{n-1}$; that is, the kind type parameter and the character length are constant.

The rank of a structure component is zero if all **part**$_k$ are scalar such as in the case of a structure component whose parent object is an array element. The rank of a structure component is greater than zero if one of the **part**$_k$ designates a whole array or an array section. In this case, the rank of the structure component is equal to the rank of the array or array section.

A structure component has the INTENT or TARGET attribute if the object **part**$_1$ has the respective attribute. A structure component is a pointer if and only if the POINTER attribute is specified in the corresponding type component **part**$_n$ of its derived type definition.

If the parent object or a type component is an array, then an array element or an array section is written by adding a subscript list or an array section subscript list, respectively, enclosed in parentheses.

```
type, public ::  measurement              ! derived type definition
   character (len=10)    :: date
   real, dimension (100) :: value
end type measurement
```

```
type, public :: old_new                  ! derived type definition
  type (measurement)     :: old
  real, dimension (100) :: factor
  type (measurement)     :: new
  character (len=20), dimension (100) :: attribute
end type old_new

type, public :: car                      ! derived type definition
  character (len=10) :: make, type
  integer :: year_of_manufacture, power
  real    :: price
end type car
```

Where these derived type definitions are available, the following derived type objects may be declared:

```
type (measurement) :: x
type (old_new)     :: z
type (car), dimension (10) :: priv
```

Example	Parent object	Type comp.	Struct. comp.
x%date	scalar	scalar	scalar
priv%price	array	scalar	array
priv(5)%price	array element, scalar	scalar	scalar
x%value	scalar	array	array
priv(1:5)%make	array section	scalar	array
z%new%date	structure comp., scalar	scalar	scalar
z%new%value	structure comp., scalar	array	array

4.6 Automatic Variables

The characteristics of a variable may depend on *specification expressions* included in its type declaration statement. These specification expressions may appear in the length specification of a character variable declaration and in the upper and lower bounds of an array declaration.

Such a variable is an **automatic variable** if it is *not* a dummy argument and if, in addition, at least one of these specification expressions is *not* a constant expression. An automatic variable can appear only in a subprogram.

Neither an initial value nor the SAVE attribute must be specified in the declaration of an automatic variable.

Automatic variables exist only for the time of the execution of the subprogram. The nonconstant character length of an automatic character variable is calculated once before the execution of the first executable statement of the subprogram. During the execution of the subprogram, the redefinition of any operand in the length specification expression does not change the length of the automatic character variable. The same length remains in effect throughout the current execution of the subprogram.

Automatic arrays are described in chapter 6.

4.7   Association

Association means that the same entity is identified in *different* scoping units by the same name or by different names. If two entities become associated, then corresponding parts of these entities also become associated.

4.7.1   Name Association

There are three ways of name association: *argument association*, *USE association*, and *host association*.

Argument association

Upon execution of a subprogram reference, that is, upon subprogram invocation, the actual arguments of the subprogram reference become associated with the dummy arguments of the referenced subprogram. These associations established by the subprogram invocation exist only for this subprogram execution. When the subprogram execution terminates, these argument associations also terminate.

USE association

By a USE statement, certain local names in a scoping unit may become associated with certain names specified within a module. Thus, entities from the module can be accessed in the scoping unit with the USE statement. This USE association remains in effect throughout the execution of the program.

Host association

A module subprogram or a derived type definition has access to entities from its host. These accessible entities are named variables, named constants, module subprograms, interface blocks, derived types, and generic identifiers.

If an entity which is accessible in the embedded scoping unit by USE association has the same nongeneric name as an entity in the host scoping unit, the entity in the host is inaccessible by that name within the embedded scoping unit.

```
module environment
  public :: inner
  type,public :: user_def
    :
  end type user_def
  integer, private :: i, j
  :
  contains
    function inner (da) result (in_var)
      real            :: da
      integer         :: in_var
      integer         :: i      ! local i
      type (user_def) :: t      ! type user_def of the host
      :
      in_var = i + j            ! the j of the host
    end function inner
end module environment
```

The derived type user_def used in the function inner is the type that is defined in the module specification part. Since there is an explicit type declaration for the variable i in the function inner, the variable i in the module specification part is *not* accessible by host association within the function.

A name that appears in the embedded scoping unit as:

- A name declared (by an interface block) to be an external subprogram name;

- A data entity name whose characteristics are described by its appearance in a type declaration statement;

- A name of a named construct; or

- A name of a subprogram, a result variable, or a dummy argument in a FUNCTION or SUBROUTINE statement

is either the name of a local entity of the scoping unit or the name of a global entity. And any entity of the host that has this as its nongeneric name is inaccessible by that name by host association. Note that entities being local to a subprogram are not accessible to its host.

4.7.2   Pointer Association

A pointer and a target may become associated by **pointer association** such that the target may be referenced or defined by actually referencing or defining the pointer.

A pointer may become associated with another target. But a pointer is associated with not more than one target at the same time. A pointer may become disassociated. A target may become associated with another pointer. A target may be associated with more than one pointer at the same time. A target may become disassociated from a pointer. The pointer association status may be *undefined*, *associated*, or *disassociated*.

4.8   Definition Status

At any time during the execution of a program, the definition status of a variable is either *defined* or *undefined*. A **defined** variable has a (valid) value. The variable remains defined with this value until it becomes either *undefined* or becomes *redefined* with a new value. A variable must be referenced only if it is defined.

If a variable is **undefined**, it has no valid value. In this case, the actual stored value of the variable is normally useless.

A variable is defined if and only if *all* its subobjects are defined. That is, an object is already undefined if only one subobject is undefined.

A pointer which is currently associated with a definable target may be defined or redefined in the same way as a "normal" nonpointer variable.

A variable (in a subroutine) may be defined already at the beginning of the execution of a program. Such an *initialized* variable has an *initial value*. Such an initial value may be explicitly specified by a type declaration statement. Each variable having an initial value must also have the SAVE attribute. No variable within the main program and within a function has an initial value.

Zero-size arrays and zero-length character variables are *always defined*, except allocatable arrays, automatic data objects, and pointers of size zero or length zero which are not defined until they are allocated or associated.

All variables which are not initialized except those that are always defined are *undefined* at the beginning of the execution of a program.

Definition status "defined"

During program execution, variables may become *defined* as a result of certain events, for instance:

The execution of an intrinsic assignment statement causes the variable on the left-hand side of the equals to become defined.

The execution of a pointer assignment with a defined target causes the pointer on the left-hand side to become defined.

The execution of an input statement causes the input list items to become defined with the values being transferred from the input file.

The execution of a DO statement with a DO variable causes this DO variable to become defined.

The execution of an input/output statement containing a control information list causes nearly all specified input/output specifiers having the character of output arguments to become defined.

The execution of a statement containing a status variable causes the status variable to become defined.

Definition status "undefined"

During program execution, variables may become *undefined* as a result of certain events, for instance:

The execution of a RETURN, END FUNCTION, or END SUBROUTINE statement in a subprogram causes all local variables of the (instance of the) subprogram to become undefined with the following exceptions:

- *Saved* variables, that is, variables with SAVE attribute, for instance, certain initialized variables in subroutines;

- Variables accessed from the host scoping unit; and

- Variables accessed from a module that is referenced directly or indirectly by at least one other scoping unit referencing the subprogram directly or indirectly.

If the execution of an input statement causes the occurrence of an error condition or an end-of-file condition, all of the variables specified in the input list of the statement become undefined.

The execution of a direct access input statement that specifies a record number for which no record has been written previously causes all of the variables specified in the input list of the statement to become undefined.

The execution of an INQUIRE statement may cause certain specifiers having the character of output arguments to become undefined.

When an allocatable array is deallocated, it becomes undefined.

Upon invocation of a subprogram, the following variables are undefined:

- An optional dummy argument that is not *present*;
- A dummy argument with INTENT(OUT) attribute;
- An actual argument associated with an INTENT(OUT) dummy argument;
- And the result variable of a function.

5 POINTERS

In F, *pointer* is not a data type, but it is an attribute which may be specified for variables or user-defined functions of any intrinsic or derived type.

5.1 Pointer Concept

A variable or a function that has the POINTER attribute is a **pointer**. This POINTER attribute must be specified in the type declaration statement.

A pointer may be *pointer associated* with a target. If a pointer is associated with a target, then the pointer *points* at this target. And the pointer may be used in place of the target wherever a data entity of the same type, kind type parameter (if applicable), character length (if applicable), and shape may be used.

A pointer shall not be defined and referenced until it is pointer associated. That is, at first the pointer must point at a target, and then this pointer may be used like any other "normal" variable. If the pointer is associated with a target, any reference to the pointer is treated as a reference to its associated target. Unlike other programming languages, the programmer need not distinguish addresses from values. Pointers are implicitly "dereferenced" in F.

There are two different possibilities of pointer association. First: when a *pointer target* is dynamically allocated (that is to say, created) by an ALLOCATE statement for a given pointer, the pointer becomes associated with this target. Second: when a *pointer assignment statement* of the form *pointer => target* is executed, the pointer becomes associated with a (new) target; we say that the pointer is "pointer-assigned" to the new target.

An *array pointer* is a named array variable having the POINTER attribute. The rank but no array bounds are specified in the array declaration for an array pointer by writing only a colon for each dimension. The array bounds may be specified later in an ALLOCATE statement. When this ALLOCATE statement is executed, the array bounds are calculated and a target array of the resulting shape is created and associated with the pointer.

```
integer, pointer :: p1
logical, dimension (:, :, :), pointer :: p2
allocate (p2 (3, 4, 15))
```

p1 is an integer pointer. And p2 is a 3-dimensional array pointer.

A type component may be specified with the POINTER attribute such that a corresponding structure component is a pointer. This way, a derived type may be defined such that its data objects may be used to form flexible data structures such as linked lists, trees, or graphs.

The ASSOCIATED intrinsic function may be used to determine whether a pointer is currently associated, disassociated, or associated with a given target.

Every target, a pointer points at, must have the TARGET attribute. The TARGET attribute must be specified in the type declaration statement.

A pointer has a definition status like any other nonpointer variable. Its definition status is that of its currently associated target.

5.2   Pointer Processing

The execution of an ALLOCATE statement for a pointer causes a target object to be created dynamically for the pointer. This target object implicitly has the TARGET attribute. The pointer and this dynamically created **pointer target** become associated and the pointer points at this target. The use of a pointer assignment statement is another possibility to associate a pointer with an(other) target or with a part of a target.

The DEALLOCATE statement may be used to *deallocate* a pointer target which was created by the execution of an ALLOCATE statement. And the NULLIFY statement may be used to *disassociate* pointers.

5.2.1   Creation of Pointer Targets

The ALLOCATE statement is used to create pointer targets.

ALLOCATE (pointer [, pointer]... [, STAT = status_variable])

Each **pointer** is a name of a scalar variable that has the POINTER attribute, the designator of a scalar structure component that has the POINTER attribute, or the name of an array pointer followed by its array specification (see chapter 6). And the **status_variable** is a scalar integer variable.

The ALLOCATE statement for pointers has the same form as the corresponding statement for array pointers (a special case, see chapter 6) and for allocatable arrays. Precisely: the same ALLOCATE statement may be used to allocate both pointer targets and allocatable arrays at the same time.

If a status variable is specified, it will become defined with the value zero when the ALLOCATE statement has been executed without an error. When an error condition occurs during the execution of the ALLOCATE statement, this status variable will become defined with a processor-dependent positive value.

If the status variable is a pointer (or an array element of an allocatable array), this variable must not be associated (or allocated) by the same ALLOCATE statement. When an error condition occurs during the execution of an ALLOCATE statement without a status variable, program execution is terminated.

It is not an error when an ALLOCATE statement is executed for a pointer which is currently associated with a target. In this case, a new pointer target is created which has all the attributes that are specified for the pointer. After successful execution of the ALLOCATE statement, the pointer points at this new pointer target and the former association is disassociated. If the former target has been created by an ALLOCATE statement, this target is no longer accessible, except there is an additional association between this pointer target and another pointer. Pointer targets which are no longer accessible are known as "dangling pointers".

5.2.2   Association Status

The association status may be *associated*, *disassociated*, or *undefined*.

Associated: a pointer receives the association status "associated" when

- The pointer is allocated as the result of the successful execution of an ALLOCATE statement referencing the pointer; or

- The pointer is pointer-assigned either to a target that is itself a currently associated pointer or to a target which has an explicitly specified TARGET attribute and which is currently allocated if it is allocatable.

Disassociated: a pointer receives the association status "disassociated" when

- The pointer is nullified by a NULLIFY statement;

- The pointer is deallocated by a DEALLOCATE statement; or

- The pointer is pointer-assigned to a currently disassociated pointer.

Undefined: a pointer receives the association status "undefined"

- When its pointer target is deallocated other than through the pointer (such as through another pointer pointing at the same target);

- When the pointer is pointer-assigned to a currently undefined pointer;

- When the execution of the END FUNCTION or END SUBROUTINE statement or a RETURN statement causes the pointer's target to become undefined; or

- During the execution of the END FUNCTION or END SUBROUTINE statement or a RETURN statement in the subprogram where the pointer was either declared or accessed unless it is one of the following:

    - A pointer with the SAVE attribute;
    - A pointer accessed from a module that is referenced directly or indirectly also by at least one other scoping unit that is referencing the subprogram either directly or indirectly;
    - A pointer accessed by host association; or
    - A pointer that is the result variable of a function with POINTER attribute.

    In case of these exceptions, the pointer retains its definition status after execution of the END FUNCTION, END SUBROUTINE, or RETURN statement. When a pointer target becomes undefined because of the execution of an END FUNCTION, END SUBROUTINE, or RETURN statement, the pointer association status also becomes undefined.

If the association status of a pointer is undefined, the pointer variable (or a subobject of it) must normally not be defined, not be referenced, and not be deallocated. Exception: such a pointer may be specified as an actual argument in a reference to any intrinsic inquiry function returning information about the pointer association status, about properties of its data type, about its kind type parameter, about its character length, or about argument presence.

If the result variable of a function is a pointer, the pointer association status of this result variable is undefined when the function is invoked. Before returning to the referencing scoping unit, this pointer must be associated with a target or the pointer must receive the pointer association status "disassociated".

If a module subprogram accesses a pointer by host association, an association which exists at the time of the subprogram reference remains existent for the time of the execution of the subprogram. The association status may be changed during the excution of the module subprogram. When the execution of the subprogram is terminated, the pointer association status remains as it is, except the currently associated target becomes undefined by the execution of the END FUNCTION, END SUBROUTINE, or RETURN statement.

5.2.3 Deallocation of Pointer Targets

A pointer target created by an ALLOCATE statement may be deallocated by the execution of a DEALLOCATE statement. **Deallocation** of a pointer target means that the association between pointer and pointer target is disassociated and that the pointer target object is no longer available.

DEALLOCATE (pointer [, pointer]... [, STAT = status_variable])

Each **pointer** is a name of a scalar variable that has the POINTER attribute, the designator of a scalar structure component that has the POINTER attribute, or the name of an array pointer (*not* followed by an array specification). And the **status_variable** is a scalar integer variable.

The DEALLOCATE statement for pointers has the same form as the corresponding statement for array pointers (a special case, see chapter 6) and for allocatable arrays. Precisely: the same DEALLOCATE statement may be used to deallocate both pointer targets and allocatable arrays at the same time.

If a status variable is specified, it will become defined with the value zero when the DEALLOCATE statement has been executed without an error. When an error condition occurs during the execution of the DALLOCATE statement, this status variable will become defined with a processor-dependent positive value. Such an error condition occurs, for example, when a DEALLOCATE statement tries to deallocate a currently disassociated pointer.

If the status variable is a pointer (or an array element of an allocatable array), this variable must not be deallocated by the same DEALLOCATE statement.

```
integer :: status_variable
⋮
deallocate (a, b, stat = status_variable)
```

A pointer must *not* be deallocated by a DEALLOCATE statement if

- The association status is undefined;
- The pointer is currently disassociated;
- The currently associated target is no pointer target which was created by the execution of an ALLOCATE statement;
- The pointer is an array pointer currently associated with an allocatable array; or
- The pointer is currently associated with a portion of a target object that is independent of any other portion of the target object.

When a pointer target is deallocated, the association status of any pointer becomes undefined that is pointing at this pointer target or that is currently associated with a subobject of this pointer target.

When an error condition occurs during the execution of a DEALLOCATE statement without a status variable, program execution is terminated.

```
subroutine sp (ip1, ip2)
  integer, pointer :: ip1, ip2
  integer :: alloc_status, dealloc_status
```

```
allocate (ip1, ip2, stat = alloc_status)
if (alloc_status > 0) then
  call alloc_error()
endif
⋮
deallocate (ip1, ip2, stat = dealloc_status)
if (dealloc_status > 0) then
  call dealloc_error()
endif
end subroutine sp
```

5.2.4   Nullification of Pointer Associations

The association between a pointer and a target may be nullified. The association between a pointer and a pointer target which was created dynamically by the execution of an ALLOCATE statement for this pointer is nullified when the pointer target is deallocated by a DEALLOCATE statement. A pointer association may also be nullified by the execution of a NULLIFY statement for the pointer.

NULLIFY (pointer [, pointer]...)

Each **pointer** is a name of a scalar variable that has the POINTER attribute, the designator of a scalar structure component that has the POINTER attribute, or the name of an array pointer (*not* followed by an array specification).

Note that (in contrary to a DEALLOCATE statement) *no* pointer target is deallocated by the execution of a NULLIFY statement.

```
complex, pointer :: cp1, cp2, cp3
allocate (cp1, cp2, cp3)
⋮
deallocate (cp1)
nullify (cp2)
cp3 => cp1
```

After execution of the NULLIFY statement, the pointer cp2 does no longer point at the allocated pointer target. The pointer target continues to exist, but it is inaccessible and cannot be deallocated. The pointer target that was allocated for cp1 is deallocted. Therefore, the pointer cp3 becomes finally disassociated, because cp3 is pointer-assigned to cp1 and cp1 is itself disassociated.

Pointer assignment statements are presented in detail in chapter 8. And array pointers are presented in chapter 6.

6 ARRAY PROCESSING

General aspects of arrays are presented in chapter 4. This chapter deals with array declaration and use of arrays.

6.1 Array Declaration

A named array must be explicitly declared by a type declaration statement to be an array. In addition to the name, each array declaration must specify the rank, and, if applicable, the shape of the array. The general form of an **array specification** is:

DIMENSION (dimension [, dimension]...)

Each **dimension** specifies the **bounds** of the array in this dimension. For every dimension, a **lower_bound** and/or an **upper_bound** may be specified, must be specified, may be omitted, or must be omitted, depending on the kind of the array.

The **lower_bound** and the **upper_bound** are special scalar integer expressions, namely *specification expressions*, which may be negative, positive, or even zero. In the list **dimension [, dimension]...** one to seven dimensions may be specified.

If **lower_bound ≤ upper_bound**, the integer values from **lower_bound** up to and including **upper_bound** determine the valid subscript values (in this dimension) for array element designators and for array section designators. But if **lower_bound > upper_bound** for a particular dimension, there are no valid subscript values in this dimension; that is, the extent of the array in this dimension is zero and, as a result, the size of the array also is zero (see below).

```
subroutine up (p)
real, dimension (5:,:), intent(in) :: p   ! assumed-shape array
real, dimension (3:8,9,7) :: q, r         ! explicit-shape arrays
real, allocatable, dimension (:,:) :: u   ! allocatable array
real, pointer, dimension (:) :: v, t      ! array pointers
:
end subroutine up
```

Variable array bounds

Array bounds that are *nonconstant* specification expressions may appear in the array declaration only of dummy arguments, of *automatic arrays*, and of result variables of array-valued functions.

In the first two cases, the array bounds and consequently the shape of the array are determined at entry to the subprogram by evaluating the array bound expressions **lower_bound** and **upper_bound**. Throughout subprogram execution, the bounds of such an array are fixed; that is, they are unaffected by any redefinition or undefinition of operands appearing in the array bounds expressions.

Zero-size arrays

An array has the size zero if its extent in (at least) one dimension is zero. Zero-size arrays are treated as "normal" arrays. Note that arrays may have different shapes even if they have the same size zero.

6.1.1   Explicit-Shape Arrays

An **explicit-shape array** is a named array with explicitly specified upper array bounds.

DIMENSION ([lower_bound :] upper_bound [, [lower_bound :] upper_bound]...)

The array specification must contain at least the **upper_bound** of every dimension. If the specification of a **lower_bound** is omitted, the default value 1 is assumed.

This form of an array specification must not be specified for dummy arguments, allocatable arrays, or array pointers, but it may appear

- In the type declaration of a local array in the specification part of a main program, a module, or a module subprogram;

- In the type declaration of an *automatic array* in the specification part of a subprogram;

- In the type declaration of a result variable for a module function; or

- In a type component definition within a derived type definition in the specification part of a module.

```
real, dimension (-5:n+1, 10, n) :: eg
integer, dimension (55, 14:22)  :: lofe
```

6.1.2   Assumed-Shape Arrays

An **assumed-shape array** is a dummy argument array without the POINTER attribute. It assumes its shape from the associated actual argument array. The array specification must contain at least a colon but no upper bound for each dimension.

DIMENSION ([lower_bound] : [, [lower_bound] :]...)

The size of a dimension of such a dummy argument array is equal to the size of the corresponding dimension of the associated actual argument array. If a **lower_bound** of a dimension is specified, the upper bound of this dimension results from the specified **lower_bound** and the *assumed* extent e of the array in this dimension as **(lower_bound** $+ e-1$). If the specification of the **lower_bound** of a dimension is omitted, the value 1 is assumed as the default lower bound.

```
real, dimension (1950:1989, 2:4) :: af
call accident(af)
⋮
subroutine accident (ff)
  real, dimension (0:, :), intent(inout) :: ff
  ⋮
end subroutine accident
```

The dummy argument array assumes the shape of the associated actual argument array. Therefore, the first dimension has the size 40 and extends from 0 to 39. And the second dimension has the size 3 and extends from 1 to 3.

6.2 Reference and Use

Whole arrays, *array sections*, and *array elements* may be used. Array sections and array elements do not have names. They are referenced by an *array section designator* and by an *array element designator*, respectively, that is the name of the parent object followed by a particular qualifier.

6.2.1 Whole Arrays

If only the name of an array is written, this is interpreted as the entire array, as all array elements, or as the starting address of the array elements in *array element order.*

Such a whole array may be a named constant or a named variable. If only the name is written, this does not mean an explicit or implicit order of the references to the single array elements. Only where particularly mentioned in this text, an automatic reference order is in effect; then the array elements are referenced in array element order.

```
real, dimension (100), save :: field = (/ 1.0, i=1,100 /)
field        = 0.0
field(33)    = 67.0
field(50:75) = (/ (-5.0, i=50,75) /)
```

The left-hand side of the second assignment statement designates an array element, the left-hand side of the third assignment statement designates an array section, and the other statements designate the whole array **field**.

6.2.2   Array Elements

An array element is a scalar part of an array entity. It does not have a name. For the reference, the name of the parent array is qualified. A simple case is an array element that is no structure component:

array (subscript_expression [, subscript_expression]...)

where **array** is the name of the parent array. The **subscript list** follows enclosed in parentheses.

Each **subscript_expression** in the subscript list is a scalar integer expression that evaluates to a valid subscript value at the time of the reference. The value of a subscript expression is valid if it is within the interval given by the specified array bounds of the dimension. The number of subscript expressions within the subscript list must correspond to the rank of the array.

```
real,dimension (3, 4) :: a
```

a(3, 2) = 17.   ←— valid

a(4, 4) = 29.   ←— not valid: the value of the first subscript is too large.

A subscript expression may have operands which are array elements or function references. Such function references must not cause **side effects**; that is, the function reference must not redefine any other operand in the subscript list.

For an array element reference, the results of the subscript expressions in the subscript list determine the array element position.

An array element has the INTENT, PARAMETER, or TARGET attribute if its parent object has the respective attribute. But an array element never has the POINTER attribute.

If a complete array is processed in such a way that the program controls the order of the processing of the single elements, then the array elements should be processed *column-wise*. This means that the array elements are processed in canonical *array element order*, which is the most efficient way to determine the position of the elements (relative to the beginning of the array). Any other order of array element references, for example row-wise processing, is allowed but is nearly always extremely inefficient.

Array element of a structure component

The form of an array element of a structure component is:

part$_1$ [**%part**$_i$]... **%part**$_n$

where **part**$_1$ is the designator of the scalar parent object of derived type. This is either the name of a (scalar) structure object or an array element designator **array (subscript_expression** [**, subscript_expression**]... **)** (see above).

Each **part**$_i$ is of derived type. It is either the name of a scalar type component of the derived type of the preceding **part**$_{i-1}$ or it is the array element designator **array (subscript_expression** [**, subscript_expression**]... **)** (see above), where **array** is the name of a type component with DIMENSION attribute of the derived type of the preceding **part**$_{i-1}$.

And **part**$_n$ is an array element designator **array (subscript_expression** [**, subscript_expression**]... **)** (see above), where **array** is the name of a type component with DIMENSION attribute of the derived type of the preceding **part**$_{n-1}$.

At least **part**$_n$ must have the "normal" form of an array element designator with a subscript list enclosed in parentheses.

```
type, public :: calendar
  integer              :: day
  character (len=10)   :: month
  integer, dimension (5) :: year
end type calendar
```

Where this derived type is available, the following type declarations may appear:

```
type (calendar), dimension (10) :: new
type (calendar) :: old
```

Then   new(5)%year(2)   is an array element of a structure component,
       old%year(2)   is an array element of a structure component, and
       new(5)%day   is a structure component but no array element.

6.2.3 Array Sections

An array section is an array object. It does not have a name. For the reference, the name of the parent array is qualified. An array section may be a part of a structure component (see below). It also may consist of substrings of array elements of the parent array.

The **extent** of an array section in a particular dimension is given by the number of valid subscript values in that dimension. The **size of an array section** is given by the product of the sizes of all dimensions of the array section.

A simple case is an array section that is no array section of a structure component, no structure component, and no array section of substrings:

array (section-subscript [, section-subscript]...)

where **array** is the name of the parent array. The **section-subscript list** follows enclosed in parentheses.

And each **section-subscript** is a *subscript expression* (see below) that defines a single subscript value, or is a *subscript-triplet* that defines a subscript value sequence, or is a *vector-subscript* that also defines a subscript value sequence. At least one of the section-subscripts must be a subscript-triplet or a vector-subscript, that is, must define a subscript value sequence.

```
real, dimension (30, 40) :: f
integer, dimension (5), parameter :: i = (/ 5, 10, 15, 20, 25 /)
```

f(1, :)	is a 1-dimensional array section consisting of all 40 elements of the first row of f.
f(10, 2:40:2)	is a 1-dimensional array section consisting of every second element of the 10th row of f.
f(1:10, 2:40:2)	is a 2-dimensional array section of shape (/ 10, 20 /) consisting of every second element of the first 10 rows of f.
f(i, 2)	is a 1-dimensional array section consisting of 5 elements of the second column of f.

A **subscript expression** is a scalar integer expression that must have a valid value at the time of the reference, except when the subscript value sequence is empty. If one of the subscript value sequences is empty, the array section has the size zero. A value of a subscript expression is valid if it is within the interval given by the specified array bounds in the corresponding dimension. The number of section-subscripts must correspond to the rank of the parent array. A subscript expression may have operands which are array elements or which are function references without side effects.

An array section consist of all those array elements of the parent array that may be determined by all possible array element subscript lists obtainable from the single subscript value and from the value sequences specified for each section subscript. The **rank of an array section** is equal to the number of subscript value sequences in the section subscript list. The order of the dimensions of an array section is given by the appearances of the subscript value sequences in the section-subscript list from the left to the right. That is, the **shape of an array**

section is a 1-dimensional array, whose ith element is the number of integer values in the sequence indicated by the ith subscript triplet or vector subscript.

An array section has the INTENT, PARAMETER, or TARGET attribute if the parent object has the respective attribute. But an array section never has the POINTER attribute.

Array section of a structure component

An array section of a structure component is not a structure component. The form the array section designator is:

part$_1$ [%part$_i$]... %part$_n$

where **part$_1$** is the designator of the parent object of derived type. It is either the name of a (scalar) structure object or an array element designator **array (subscript_expression [, subscript_expression]...)**.

Each **part$_i$** is of derived type. It is either the name of a type component without the DIMENSION attribute of the preceding **part$_{i-1}$**, or it is an array element designator **array (subscript_expression [, subscript_expression]...)**, where **array** is the name of a type component with DIMENSION attribute of the preceding **part$_{i-1}$**.

And **part$_n$** is an array section designator **array (section-subscript [, section-subscript]...)** (see above), where **array** is the name of a type component with DIMENSION attribute of the preceding **part$_{n-1}$**. Only **part$_n$** designates an array object.

With regard to rank and shape of such an array section, the rules of the last section hold.

Using the derived type `calendar` and the type declarations at the end of section 6.2.2, then

`old%year(:2)`   is an array section of a structure component, and
`new(7)%year(:)`   is an array section of a structure component.

Array section is structure component

If an array section is a structure component, the form of the array section designator is:

part$_1$ [%part$_i$]... %part$_n$

where **part$_1$** is the designator of the parent object of derived type. It is either the **name** of a structure object or of an array object of derived type, or it is an array element designator **array (subscript_expression [, subscript_expression]...)**, where **array** is the name of an array of derived type, or it is an array section designator **array (section-subscript [, section-subscript]...)** (see above), where **array** is the name of an array of derived type.

Each **part**$_i$ is of derived type. It is either the **component_name** of a type component of the type of the preceding **part**$_{i-1}$, or it is an array element designator **array (subscript_expression [, subscript_expression]...)**, where **array** is the name of a type component with DIMENSION attribute of the preceding **part**$_{i-1}$, or it is an array section designator **array (section-subscript [, section-subscript]...)** (see above), where **array** is the name of a type component with DIMENSION attribute of the preceding **part**$_{i-1}$.

And **part**$_n$ is the name of a type component of the type of the preceding **part**$_{n-1}$. Exactly one of these **parts** must designate a whole array or an array section.

With regard to type, kind type parameter (if applicable), character length (if applicable), and attributes, all the normal rules for structure components hold. If such an array section is written with section-subscripts, the rank and the shape of the array section is given as for the simple form of an array section described earlier. If such an array section is written without a section-subscript, the rank and the shape of the array section is equal to the rank and shape of the part **name**, or **component_name**, or **part**$_n$, respectively, that designates the array object.

Using the derived type `calendar` and the type declarations at the end of section 6.2.2, then

`new(5:10)%day` is an array section that is a structure component,
`new(2)%year` is an array section that is a structure component, and
`old%year(:)` is an array section that is *no* structure component.

6.2.3.1 Subscript-Triplet

A section-subscript defining a subscript value sequence may be written as a *vector-subscript* or as a *subscript-triplet*. The form of a **subscript-triplet** is:

[**first**] : [**last**] [: **stride**]

where **first** and **last** are *subscript expressions* (see above), specifying the first value and the upper limit, respectively, of the subscript value sequence. If **first** is omitted, the lower bound of the dimension of the parent array is assumed as the default value. The absolute value of **last** is greater than or equal to the last value of the subscript value sequence. If **last** is omitted, the upper bound of the dimension of the parent array is assumed as the default value. The **stride** is a scalar nonzero integer expression specifying the stepsize of the subscript values of the subscript value sequence. If the **stride** is omitted, the value 1 is assumed as the default stride.

The **extent of an array section** in a dimension is given by the number of valid subscript values in this dimension, which is calculated as

$$\text{MAX} (\text{INT} ((\textbf{last} - \textbf{first} + \textbf{stride}) \,/\, \textbf{stride}, 0)) .$$

If **stride** > 0, the subscript value sequence contains the values **first** $+ n*$**stride** for $n = 0, 1, 2, \ldots$ up to the upper limit **last**. The subscript value sequence is *empty* if **first** $>$ **last**.

If **stride** < 0, the subscript value sequence contains the values **first** $- n*$**stride** for $n = 0, 1, 2, \ldots$ down to lower limit **last**. The subscript value sequence is *empty* if **first** $<$ **last**.

Note that the values of the subscript expressions of a subscript-triplet need not be valid subscript values. But the selected array elements must be within the specified array bounds.

Subscript-triplet	Subscript values
`5 : 12`	5, 6, 7, 8, 9, 10, 11, and 12.
`5 : 12 : 3`	5, 8, and 11.
`23 : 3 : 2`	the subscript value sequence is empty.
`4 : -3 : -1`	4, 3, 2, 1, 0, -1, -2, and -3.
`4 : -4 : -3`	4, 1, and -2.
`10 : 990 : -2`	the subscript value sequence is empty.

Suppose, array **f** is declared as:

`real, dimension (-20 : 1000) :: f`

Then the following array sections consist of the following array elements of the parent array **f**:

Array section	Array elements in array element order
`f(5 : 12)`	f(5), f(6), f(7), f(8), f(9), f(10), f(11), and f(12).
`f(5 : 12 : 3)`	f(5), f(8), and f(11).
`f(4 : -3 : -1)`	f(4), f(3), f(2), f(1), f(0), f(-1), f(-2), and f(-3).
`f(4 : -3 : -2)`	f(4), f(2), f(0), and f(-2).
`f(10 : 990 : -2)`	array section of size zero.

`real, dimension (4, 3, 5) :: a`

`a(3, 2, :)`   is a 1-dimensional array section that has the shape $(/\,5\,/)$. Its size is 5.

`a(:, 3, :)`   is a 2-dimensional array section that has the shape $(/\,4,\,5\,/)$. It consists of all array elements of the second columns of all 5 planes of array **a**. Its size is 20.

`a(2:4, :, 3:5)`   is a 3-dimensional array section that has the shape $(/ 3, 3, 3 /)$. Its size is 27.

`a(3, 3, 5:)`   is a 1-dimensional array section that has the shape $(/ 1 /)$.

The selected array elements of an array section are chained in array element order. In this case, the array element order is the order defined for the subobject. The array element order of an array section has nothing to do with the array element order of its parent array.

```
real, dimension (4, 3) :: r     ! <-- parent array
```

The following array section of the parent array r has the same shape as the parent array. It consists of *all* array elements of array r, but the array element order of the new array object is different from that of its parent array r.

Array section: `r(4:1:-1, 3:1:-1)`

	Column 1	Column 2	Column 3
Row 1	r(4,3)	r(4,2)	r(4,1)
Row 2	r(3,3)	r(3,2)	r(3,1)
Row 3	r(2,3)	r(2,2)	r(2,1)
Row 4	r(1,3)	r(1,2)	r(1,1)

The order of the array elements of the array section is as follows:

Array element	r(4,3)	r(3,3)	r(2,3)	r(1,3)	r(4,2)	r(3,2)	$\cdots$
Position	1	2	3	4	5	6	$\cdots$

Array element	r(2,2)	r(1,2)	r(4,1)	r(3,1)	r(2,1)	r(1,1)
Position	7	8	9	10	11	12

```
real, dimension (3, 3, 2) :: s     ! <-- parent array
```

Array section: `s(3:1:-2, 1:3:2, 1:2)`

The array section has the following form:

	Column 1	Column 2
Row 1	s(3,1,1)	s(3,3,1)
Row 2	s(1,1,1)	s(1,3,1)

Plane 1

	Column 1	Column 2
	s(3,1,2)	s(3,3,2)
	s(1,1,2)	s(1,3,2)

Plane 2

The order of the array elements of the array section is as follows:

Array elem.	s(3,1,1)	s(1,1,1)	s(3,3,1)	s(1,3,1)	s(3,1,2)	s(1,1,2)	s(3,3,2)	s(1,3,2)
Position	1	2	.3	4	5	6	7	8

6.2.3.2 Vector-Subscript

A section-subscript defining a subscript value sequence may be written as a *subscript-triplet* or as a *vector-subscript*. A **vector-subscript** is a 1-dimensional integer array expression. Every array element of such a (vector-subscript) array expression must be defined with a valid subscript value.

```
integer, dimension (3) :: u
integer, dimension (4) :: v
real, dimension (5, 7) :: z
```

Suppose, arrays u and v are defined as follows:

```
u = (/1, 3, 2/)
v = (/2, 1, 4, 3/)
```

Then the array section z(3, v) is a 1-dimensional array section consisting of the following elements (in this order):

   z(3, 2)   z(3, 1)   z(3, 4)   z(3, 3)

And the array section z(u, 2) is a 1-dimensional array section consisting of the following array elements (in this order):

   z(1, 2)   z(3, 2)   z(2, 2)

And the array section z(u, v) is a 2-dimensional array section consisting of the following array elements:

	Column 1	Column 2	Column 3	Column 4
Row 1	z(1,2)	z(1,1)	z(1,4)	z(1,3)
Row 2	z(3,2)	z(3,1)	z(3,4)	z(3,3)
Row 3	z(2,2)	z(2,1)	z(2,4)	z(2,3)

Array element	z(1,2)	z(3,2)	z(2,2)	z(1,1)	z(3,1)	z(2,1)	···
Position	1	2	3	4	5	6	···

Array element	z(1,4)	z(3,4)	z(2,4)	z(1,3)	z(3,3)	z(2,3)
Position	7	8	9	10	11	12

Array sections *with* vector-subscripts are subject to certain restrictions which are normally not effective for array sections *without* vector-subscripts. There are restrictions on their use as internal files, as actual arguments, as targets in pointer-assignment statements, and on their use on the left-hand side of intrinsic assignment statements. ·

6.2.3.3   Array Sections of Substrings

If an array section is of type character, a character string notation may be written immediately after the "normal" array section designator. This designates a subobject of an array section that has the same shape as the array section; and each element of the subobject consists only of the specified character positions of the array elements of the array section. Such a subobject of an array (section) also is an array section, but it is no character string.

```
character (len=10), dimension (5, 10, 10) :: z
```

Then `z(:, :, 4)(6:10)` is a 2-dimensional array section that has the shape (/ 5, 10 /). Its array elements are character substrings of the corresponding elements of its parent array `z` and have the character length 5.

6.3   Memory Management and Dynamic Control

If a subprogram needs an array which is no dummy argument array and which has a shape that depends on one or more variables, then the programmer may use an *automatic array*. And if, for some time, a local array is needed that has a shape which must be dynamically calculated before the array is allocated, then the programmer may use an *allocatable array* or an *array pointer*.

6.3.1   Automatic Arrays

An **automatic array** may be declared only in the scoping unit of a subprogram. It is no dummy argument array but an explicit-shape array which has at least one array bound that is a *nonconstant* specification expression.

The lifetime of an automatic array is limited. It is automatically created (allocated) on entry to the subprogram and is automatically deallocated on return to the referencing scoping unit.

```
subroutine exchange (f1, f2)
  real, dimension (:, :), intent(inout)     :: f1, f2
  real, dimension (size(f1,1), size(f1,2)) :: hf
  hf = f1
  f1 = f2
  f2 = hf
end subroutine exchange
```

The automatic array **hf** is allocated when the subroutine **exchange** is invoked, and it is deallocated when the END SUBROUTINE statement of the subroutine is executed. The programmer need not take care of allocation and deallocation operations.

An automatic array *exists* only for the execution of the (instance of the) subprogram. Therefore, an automatic array must not be specified to have the SAVE or PARAMETER attribute. As long as an automatic array exists, that is, throughout the execution of the (instance of the) subprogram, a redefinition of any operand in one of the nonconstant bounds expressions has no effect on the current shape of the automatic array.

6.3.2 Allocatable Arrays

An **allocatable array** is a named array (but *no* dummy argument array and *no* function result) for which the ALLOCATABLE attribute is specified in its type declaration statement. The array declaration contains only a colon for each dimension but no array bounds. The form of such an array specification is:

DIMENSION (: [, :]...)

The explicit shape for an allocatable array is specified in an ALLOCATE statement. And the (storage for the) array is allocated when this ALLOCATE statement is executed; that is, the array becomes dynamically allocated. The allocatable array is available until it is automatically deallocated or until a DEALLOCATE statement is executed for this array.

Type, kind type parameter (if applicable), character length (if applicable), name, rank, and the ALLOCATABLE attribute of an allocatable array must be specified in the specification part of the scoping unit containing the ALLOCATE statement or must be accessible from the host or from a module. Only the array bounds of all dimensions remain to be determined during the execution of the program.

The rank of an allocatable array is given by the number of colons in its array declaration. The TARGET attribute is allowed to be specified for an allocatable array; that is, a pointer may point at an allocatable array.

```
real, dimension (:, :), allocatable        :: a
integer, allocatable, dimension (:, :, :) :: b
```

a is a 2-dimensional and b is a 3-dimensional allocatable array.

As long as no ALLOCATE statement is executed for an allocatable array, that is, as long as it is not allocated, its array bounds, its shape, and its size are undefined.

Only the lower and upper bounds of all dimensions are dynamically determined by the execution of the ALLOCATE statement for the array. Thus the extents in all dimensions, the shape, and the size of the array become defined.

ALLOCATE (array (dim [, dim]...) [, array (dim [, dim]...)]...
[, STAT = status_variable])

Each **array** is the name of an allocatable array available in the scoping unit containing the ALLOCATE statement. **(dim [, dim]...)** is an array specification, which has the form as for an explicit-shape array. And the **status_variable** is a scalar integer variable. The array bounds expressions must be scalar integer expressions; they need not be specification expressions.

The ALLOCATE statement for allocatable arrays has the same form as the corresponding statement for pointers. Precisely: one ALLOCATE statement may be used to allocate allocatable arrays and pointer targets at the same time. See chapter 5 for the use of the status variable.

An allocatable array is a "normal" local array but never a structure component. Its allocated size may be zero. And it may be a zero-length character array.

When an ALLOCATE statement is executed, the array bounds are determined for each dimension, and then the (storage for the) array is allocated. As long as the allocatable array remains allocated, the redefinition of any operand in its array bounds expressions has no effect on the current array bounds of the allocatable array. Its array bounds expressions may become even undefined.

```
integer :: alloc_error
real, allocatable, dimension (:, :)     :: a
real, allocatable, dimension (:, :, :) :: b
⋮
allocate (a (-n:-1, n), b (25:75, 2, n), stat=alloc_error)
```

Allocation status: An allocatable array becomes **allocated** after successful execution of an ALLOCATE statement until it becomes *deallocated*.

A **deallocated** allocatable array is no longer allocated. If an allocatable array is currently not allocated, either no ALLOCATE statement has been (successfully) executed or the array is already deallocated. Only a currently allocated allocatable array may be defined and referenced.

By execution of the END FUNCTION or END SUBROUTINE statement or a RETURN statement in a subprogram, the following currently allocated allocatable arrays in the subprogram retain their allocation status and their definition status:

- An allocatable array with SAVE attribute;

- An allocatable array which is accessible by host association; or

- An allocatable array in the scoping unit of a module which also is accessed by another scoping unit that is currently in execution.

Deallocation: A currently allocated allocatable array may be deallocated by the execution of a DEALLOCATE statement. **Deallocation** of an allocatable array means that the storage allocated for the array is no more available.

DEALLOCATE (array [, array]... [, STAT = status_variable])

The DEALLOCATE statement for allocatable arrays has the same form as the corresponding statement for pointers. Precisely: one DEALLOCATE statement may be used to deallocate allocatable arrays and pointer targets at the same time. See chapter 5 for the use of the status variable.

```
integer :: dealloc_stat
:
deallocate (a, b, stat = dealloc_stat)
```

An allocatable array with TARGET attribute must not be deallocated by using an associated pointer. If such an allocatable array is deallocated, all currently associated pointers become undefined. An allocatable array with undefined allocation status must not be deallocated.

```
subroutine ff (ja, je)
  integer, intent(in) :: ja, je
  integer             :: istat
  real, allocatable, dimension (:, :) :: df1, df2
  allocate (df1 (ja:je, 7), df2 (1950:je, 7), stat= istat)
  if (istat > 0) then
    call al_error()
  endif
  call compare(1950, ja, je, df1, df2)
  deallocate (df1, df2, stat= istat)
  if (istat > 0) then
    call deal_error()
  endif
end subroutine ff
```

6.3.3   Array Pointers

An **array pointer** is a named array for which the POINTER attribute is specified in its type declaration statement. The array declaration contains only
a colon for each dimension but no array bounds. The form of such an array
specification is:

DIMENSION (: [, :]...)

Type, kind type parameter (if applicable), character length (if applicable), name,
rank, and the POINTER attribute of an array pointer must be specified in its
type declaration statement. Only the array bounds of all dimensions remain to
be determined during the execution of the program. The character length of an
array pointer may be zero. The rank of an array pointer is given by the number
of colons in the array declaration.

During the execution of the program, the array bounds and the shape are determined either by the execution of an ALLOCATE statement, which specifies the
explicit shape of the pointer target, or by the execution of a pointer assignment
statement, which associates a target with the array pointer.

When an ALLOCATE statement is executed for the array pointer, the array
bounds are determined for each dimension, and then the pointer target array
is allocated (created). This dynamically created pointer target (implicitly) has
the TARGET attribute. The array pointer and this pointer target array become
pointer associated, and the array pointer may be used, as any other pointer, to
reference and to define this pointer target. By the execution of pointer assignment statements, additional pointers may be associated with this pointer target
array or with parts of it. In this case, two or more pointers are pointing at the
same target.

```
real, dimension (:, :), pointer      :: c
integer, pointer, dimension (:, :, :) :: d
```

c ist a 2-dimensional and d is a 3-dimensional array pointer.

The lower bound and the upper bound of all dimensions are dynamically determined by the execution of the ALLOCATE statement for the array pointer.
Thus the extents of the array in all dimensions, the shape of the pointer target
array, and its size become defined.

The ALLOCATE statement for array pointers has the same form as the corresponding statement for allocatable arrays (see last section).

The ALLOCATE statement must specify the array bounds of *all* dimensions of
an array pointer. The array bounds expressions must be scalar integer expressions; they need not be specification expressions. The size of the allocated target
array may be zero.

```
integer :: alloc_error
real, pointer, dimension (:, :)          :: a
complex, pointer, dimension (:, :, :) :: b
allocate (a (-n:-1, n), b (25:75, 2, n), stat = alloc_error)
```

If the array bounds of an array pointer are determined such that the array pointer *ap* becomes associated with a target *t* by pointer-assignment, such as *ap* => *t*, then the lower bounds are given by the function result of LBOUND(*t*) and the upper bounds by the result of UBOUND(*t*).

Deallocation of pointer targets: The pointer target of an array pointer **6** allocated with an ALLOCATE statement may be deallocated (like any other target of this kind) by the execution of an DEALLOCATE statement; see chapter 5 for exceptions.

```
subroutine pp (ja, je)
  integer, intent(in) :: ja, je
  integer :: istat
  real, pointer, dimension (:, :) :: ap1, ap2
  allocate (ap1 (ja:je, 7), ap2 (1950:je, 7), stat = istat)
  if (istat > 0) then
    call al_err()
  endif
  call compare(1950, ja, je, ap1, ap2)
  deallocate (ap1, ap2, stat=istat)
  if (istat > 0) then
    call deal_err()
  endif
end subroutine pp
```

Nullification: The association between an array pointer and a target array may be nullified like any other pointer association (see chapter 5).

6.4   Array Constructor

An **array constructor** is a sequence of scalar values which is interpreted as a 1-dimensional array. The scalar values in this sequence are given by a value-list consisting of single scalar values, array values, and/or implied-DOs.

```
integer, dimension (31)    :: x
integer, dimension (3, 4) :: f
:
x = (/ 0, 1, x(11) - 2, 3, 4, (i+5, 0, i=5,10), 11, 12, f /)
```

An array constructor has a type, (if applicable) a kind type parameter, and (if applicable) a character length, but it does not have a name (except when it is a named array constant). Type, kind type parameter, and length of the array constructor correspond to those of its elements. Therefore, all values in the value-list must have the same type, (if applicable) the same kind type parameter, and (if applicable) the same character length.

(/ value-list /)

The **value-list** may be empty. Each value of the list may be a scalar expression, an array expression, or the following *implied-DO*:

(value-list, do_variable = first, last [, stride])

The **do_variable** is the name of a scalar integer variable which must not be a dummy argument, must not be a pointer, must not be a function result, must not be accessed by USE or host association, must not be initialized, and must not have the SAVE attribute. And **first, last,** and **stride** are scalar integer expressions. If the stride is omitted, the value 1 is assumed as the default stride.

Array expression: An array expression in the value-list is interpreted as a sequence of scalar values in array element order.

Implied-DO: An implied-DO in the value-list is interpreted as a sequence of expressions which is expanded under control of the DO variable. Loop control is the same as for a "normal" DO construct. The value-list in an implied-DO may contain another implied-DO; that is, implied-DOs may be nested. Note that each of the implied-DOs in such a nest must have its own unambiguous DO variable.

```
real, dimension (28)     :: x, y
logical, dimension (7:5) :: l
integer, dimension (5:7) :: v
integer, dimension (77)  :: z
 :
l = (/ /)
v = (/ 5+1, 6, 27-3 /)
x = (/ y(10:14) + y(20:24), sqrt(r), (100.0/i, i=5,25), 99.0 /)
z = (/ (((i, i=1,5), j=1,4), k=1,3), 0, ((i, i=9,6,-1), j=1,4) /)
```

Empty value list: If the value list of an array constructor is empty, this is interpreted as a (1-dimensional) array of size zero.

Constant array constructor

The value-list of a **constant array constructor** contains only *constant expressions*. A constant array expression may be named.

```
integer, dimension (5) :: x
integer, dimension (7) :: y
integer, dimension (6), parameter :: a = (/ 0, 0, 0, 1, 1, 1 /)
character (len=5), dimension (3)  :: name
x    = (/ 1, 3, 5, 7, 9 /)           ! or  x = (/ (1, l=1,10,2) /)
y    = (/ 1, 1, 1, 1, 1, 1, 1 /)     ! or  y = (/ (1, k=1,7) /)
name = (/"Peter", "Jerry", "Brian"/) ! all the same length
```

6.5 Operations on Arrays

The numeric, relational, logical, and character intrinsic operators, the intrinsic assignments, and most intrinsic functions (namely the elemental functions) are defined for scalar operands and arguments, respectively, but they may also be applied element-by-element to array operands and arguments, respectively.

6.5.1 Array Expressions

An **array operand** may be a whole array, an array section, an array constructor, an array function reference, or an array expression enclosed in parentheses.

Two arrays are **conformable** if they have the same shape, that is, if they have the same rank and if the corresponding dimensions have the same size. A scalar entity is per definition conformable with every array. Such a scalar may be interpreted as a conformable array with all array elements having the same value as the scalar.

```
real, dimension (5, 3)    :: a, d
real, dimension (8, 5)    :: b
real, dimension (5, 3, 4) :: c
```

Conformable arrays or array sections are:

a	and	d
a(:, 3)	and	b(4:8:1, 1)
a(3:5, 2)	and	b(2, 3:5)
a(1:3, 1)	and	c(2, :, 4)
a	and	b(1:5, 1:3)
a	and	b(4:8, 3:5)
a(1:4, :)	and	c(2:5, 2, 2:4)

If the binary *intrinsic* operators +, −, *, ... are applied to array operands, the operands must be conformable. Such operations are performed element-by-element and produce a resultant array conformable with the array operand(s).

In this case, corresponding elements of the operands are involved as in a scalar-like operation and produce the corresponding element in the conformable array-valued result. These elemental operations can be performed in any order or even simultaneously. Such an operator is called an **elemental operator**.

```
real, dimension (3, 3) :: a
real, dimension (9, 3) :: b
real, dimension (3, 9) :: c
```

Array expressions are:

```
a + b(3:9:3, :)
a(:, 1) * a(:, 2) / ( a(:, 3) - a(:, 1) )
c(1, 7:9) ** a(3, 3)
( b(2:3, 1:2) + 4.7e11 ) * ( c(1:2, 5:6) - 0.8e15 )
```

```
integer, dimension (3, 4, 5) :: a, b
real, dimension (3, 4, 5)    :: c, d
:
c = a * b
d = d + 5
```

The right-hand side of the first assignment is no matrix multiplication, but it is an elemental multiplication of two arrays. The resultant array has the same shape as the operand arrays. In the second assignment, the value 5 is added to *each* element in array d.

6.5.2   Array Subprograms

The programmer may use elemental intrinsic subprograms which are defined for scalar arguments but which may also be referenced with array arguments. He may write his own array-valued functions and he may use a lot of array intrinsic functions.

There are intrinsic functions for

- Matrix multiplication, vector-matrix multiplication (MATMUL) and dot-product (DOT_PRODUCT);

- Numeric and logical computations with reduction of the rank of the result (SUM, PRODUCT, MAXVAL, MINVAL, COUNT, ANY, and ALL);

- Inquiry of certain characteristics of arrays (ALLOCATED, SIZE, SHAPE, LBOUND, and UBOUND);

- Construction of arrays (MERGE, SPREAD, RESHAPE, PACK, and UNPACK);

- Manipulation (such as transposition) of arrays (TRANSPOSE, EOSHIFT, and CSHIFT); and

- Determination of the position of certain array elements (MAXLOC and MINLOC).

6.5.3   Array Assignments

If the left-hand side of an intrinsic assignment statement is an array, this statement is an **array assignment statement**.

If the left-hand side is an array and the right-hand side is scalar, all elements
of the array on the left-hand side are (re)defined with the same value of the
right-hand side. If the right-hand side also is an array entity, both sides of the
equals must be conformable array entities.

An array assignment operation is *elemental*. That is, the left-hand side of an
array assignment statement is defined element-wise for corresponding array elements; the first element of the left-hand side becomes defined with the first
element of the right-hand side, the second element of the left-hand side becomes
defined with the second element of the right-hand side, and so on. There is no
order given for the processing of these elemental assignments; they may be even
processed in parallel.

```
logical, dimension (9, 5, 3) :: x
logical, dimension (9, 3)    :: y
logical, dimension (3, 3)    :: z

z = x(1:5:2, 2, :) .or. y(9:5:-2, :)
```

is equivalent with the following scalar assignment statements in any order:

```
z(1, 1)  =  x(1, 2, 1) .or. y(9, 1)
z(2, 1)  =  x(3, 2, 1) .or. y(7, 1)
z(3, 1)  =  x(5, 2, 1) .or. y(5, 1)
z(1, 2)  =  x(1, 2, 2) .or. y(9, 2)
z(2, 2)  =  x(3, 2, 2) .or. y(7, 2)
z(3, 2)  =  x(5, 2, 2) .or. y(5, 2)
z(1, 3)  =  x(1, 2, 3) .or. y(9, 3)
z(2, 3)  =  x(3, 2, 3) .or. y(7, 3)
z(3, 3)  =  x(5, 2, 3) .or. y(5, 3)
```

The actual assignment (i. e., storing) of the result of the right-hand side is done
not before all computations on the right-hand side are done and not before all
address computations for array elements or other subobjects on the left-hand
side are done.

The evaluation of the right-hand side must neither affect nor be affected by the computations for the reference to the left-hand side; side effects are not allowed.

```
real, dimension (n) :: a, b, c
  :
a(2:n) = a(1:n-1) * b(2:n) + c(2:n)
  :
do i=2,n
  a(i) = a(i-1) * b(i) + c(i)
enddo
```

The array assignment statement and the DO construct are not equivalent because the DO construct contains data dependencies.

When array assignments are to be evaluated not for all but only for particular array elements, a *WHERE construct* may be used (see chapter 8).

7 EXPRESSIONS

An **expression** is a formula for the computation of a value. It consists of operands and operators. **Operands** are constants, variables, array constructors, structure constructors, function references, and (sub)expressions enclosed in parentheses. An operand is a scalar or an array.

```
3.1415    .true.    "Berlin"    ← scalar (literal) constants
(/7,13,24,6,8/)  (/1,0,0,1/)    ← array constructors
x      dor      week      name  ← scalar named variables, whole arrays
dor(4)    t(10)    week(14)      ← array elements
dor(1:3)          week(2:19)     ← array sections
address%po      address%cip     ← structure components
name(:5)        name(11:21)     ← character substrings
sin(x)   log(x=10)  f(dor,2)    ← function references
(r*sin(x/y) + pi/2 + 55.571)   ← subexpression
```

7

Operators designate the computations (operations). There are *binary* operators processing two operands, and *unary* operators processing one operand. Operators are either *intrinsic* or *defined*. Intrinsic operators are predefined and may be used in all parts of the program. Defined operators must be defined by the programmer and can be used only in the scoping unit where the operator is defined or where its definition is accessible.

Intrinsic operators:

Numeric operators:	$+$	$-$	$*$	$/$	$**$	
Relational operators:	$>$	$>=$	$<$	$<=$	$==$	$/=$
Logical operators:	.NOT.	.AND.	.OR.	.EQV.	.NEQV.	
Character operator:	$//$					

An expression evaluates to a value, which has a type, (if applicable) a kind type parameter, (if applicable) a character length, and a shape. It is scalar or array-valued.

Two arrays are **conformable** if they have the same shape. A scalar is by definition conformable with every array.

Elemental operators are defined for scalar operands but may also be applied to conformable operands. The intrinsic operators are predefined as elemental operators. When an array operand is being processed by such an operator, the operation is performed element-by-element without any given order thus that each array element of the result has the same array element position as the corresponding processed array element of the operand(s). For an unary elemental operator, the result of the element-wise operations is an array with the same shape as the operand. For a binary elemental operator with two array operands, the operands must be in shape conformance und the result of the element-wise operations is an array with the same shape as the operands. For a binary elemental operator with one scalar operand and one (non-zero size) array operand, the scalar operand is treated as if it were expanded into an array whose shape is that of the array operand and whose array element values are equal to that of the scalar operand.

An expression is not a statement, but it is a part of a statement. There are different kinds of expressions:

- Predefined expressions:
  - numeric intrinsic expressions, such as   `(a + b) * sin(c)`
  - relational intrinsic expressions, such as   `d <= e`
  - logical intrinsic expressions, such as   `f .and. (g .or. h)`
  - character intrinsic expressions, such as   `i // j // "klm"`

- Defined expressions, such as   `l .plus. (m .rest. line)`

Numeric intrinsic expressions, relational intrinsic expressions, logical intrinsic expressions, and character intrinsic expressions process operands of intrinsic types by intrinsic operators. Thus, a numeric intrinsic expression evaluates to a numeric result, a logical intrinsic expression evaluates to a result of type logical, and a character intrinsic expression evaluates to a result of type character. A relational intrinsic expression evaluates to a logical result and may be used as an operand only in a logical expression.

Defined expressions also may contain operands of derived types which may be processed by defined operators or even by extended intrinsic operators. A defined expression also may be a numeric, logical, or character expression if it evaluates to a result of such an intrinsic type.

An expression is either a *scalar expression* or an *array expression*. A **scalar expression** contains only scalar (primary) operands. An **array expression** has at least one (primary) *array operand*. An **array operand** is a whole array, an array section, an array constructor, a reference to an array-valued function, or an array expression enclosed in parentheses.

Constant expressions, initialization expressions and *specification expressions* are special purpose expressions mainly appearing in specification statements. They are presented at the end of this chapter.

Interpretation

At first, an F processor must establish the interpretation of an expression on the basis of the *interpretation rules*. Then the F processor may actually evaluate an *equivalent* expression instead of the given expression. A parenthesized expression is treated as a data entity by the F processor; that is, subexpressions enclosed in parentheses are invariant against internal equivalence transformations.

Two numeric intrinsic expressions are (mathematically) **equivalent** if their mathematical results — in accordance with the F rules — are equal for all possible operands. Two relational or logical intrinsic expressions are (logically) **equivalent** if their logical results are equal for all possible operands.

7.1   Numeric Intrinsic Expressions

A *numeric expression* is used to express numeric computations. A numeric expression evaluates to a scalar numeric value or to a numeric array value. A **numeric value** is an integer, real, or complex value. A **numeric intrinsic expression** consists of *numeric operands* and *numeric operators*.

```
a + b * c - (3.5 + x) ** 2 / y + sin(z)
```

Numeric operands are numeric constants, numeric variables, numeric array constructors, high precedence numeric defined subexpressions (see 7.6.1), numeric function references, and (intrinsic and defined) (sub)expressions enclosed in parentheses.

Simple numeric expressions may consist only of one operand without an operator:

```
x    10.3    0    ijk_1    str7    1    cos(y)    ff(1:3, 985:988)
```

More complicated numeric expressions consist of one or more operands which are processed by *numeric intrinsic operators*. They may contain numeric subexpressions enclosed in parentheses. These parentheses must appear in pairs.

Numeric intrinsic operators:

Operator	Operation	Use	Interpretation
**	Exponentiation	x ** y	Raise x to the power y
*	Multiplication	x * y	Multiply x by y
/	Division	x / y	Divide x by y
+	Addition	x + y	Add x and y
+	Identity	+ x	Same as x (without a sign)
−	Subtraction	x - y	Subtract y from x
−	Numeric Negation	- x	Negate x

Except for the division, the interpretation of the numeric intrinsic operators is the same as the interpretation of the corresponding algebraic operators for scalar operands. The interpretation of the division depends on the data types of the operands (see below).

Precedence of operators: The precedence of the operators is the same as in the algebra: the operator ** takes precedence over *, /, +, and −. And the operators * and / take precedence over + and −.

Interpretation rules: If a numeric intrinsic expression consists of several operands and operators, the expression is interpreted in accordance with the following rules:

- A numeric intrinsic expression is interpreted from the left to the right. That is, if there is an operand between two operators of the same precedence (except exponentiation), the left operator is combined with the operand.

- A parenthesized subexpression is treated as a data entity.

- If an expression includes operators of different precedence, the precedence of the operators controls the order of the combination of operators and operands.

- A sequence of exponentiations is combined from the right to the left; for example, 2**3**4 is interpreted as $2^{(3^4)}$.

5 - a ** 2 + b * 3   is interpreted as $(5 - (a^2)) + (b * 3)$

because exponentiation has precedence over subtraction and because multiplication has precedence over addition.

A numeric expression must not contain consecutive numeric operators:

```
a + - b      a ** - 2     ! not allowed
a + (-b)     a ** (-2)    ! allowed
```

Type, kind type parameter, and shape

An integer, real, or complex expression evaluates to a result that is an integer, real, or complex value, respectively. Type, kind type parameter, shape, and value of a numeric intrinsic expression result from those of the operand(s) and from the interpretation of the expression.

Unary numeric operator: If the operator + (plus) or − (minus) is applied to one numeric operand, the type of the expression (and thus the type of the result) is the same as the type of the operand. There are similar rules for the kind type parameters and for the shapes of the operand and of the result.

Binary numeric operator: If a numeric operator is applied to two operands of the same numeric type, the type of the expression (and thus the type of the result) is the same as the type of the operands. That is, if all operands of a numeric expression are of the same type, the result of the expression also is of this type. And if all its operands also have the same kind type parameter and the same shape, the result of the expression also has this kind type parameter and this shape.

A binary numeric operator may be applied to two operands of different shapes (if one operand is scalar), two operands of the same numeric type but with different kind type parameters, or two numeric operands of different numeric types.

Precedence of numeric types: If at least two operands in a numeric intrinsic expression are of different types, the precedence of the numeric types determines the data type of the (result of the) expression:

Type	Precedence
complex	high
real	middle
integer	low

7

`7.3 + 17 - 2.2e-4   ! is a real expression.`

Type conversion: When a binary numeric operator is applied to two operands of different numeric types, first the value of the operand with the lower type precedence is internally converted to the type of the operand with the higher type precedence, then the operator is applied to these two values of the same type. The result of the operation has the type of the operand with the higher type precedence. The only exception to this rule is the exponentiation where a real or complex value is raised to an integer power.

Kind type parameter conversion: When a binary numeric operator is applied to two integer operands with different kind type parameters, first the value of the operand with the smaller decimal range is internally converted to the kind type parameter of the other operand, then the operator is applied to these two integer values having the same kind type parameter. The result of the operation has the same kind type parameter as the operand with the greater range. The only exception to this rule is the exponentiation where an integer value is raised to an integer power.

When a binary numeric operator is applied to two real operands, two complex operands, or one real and one complex operand with different kind type parameters, first the value of the operand with the smaller decimal precision is internally converted to the kind type parameter of the other operand, then the operator is applied to these two real or two complex values having the same kind type parameter. The result of the operation has the same kind type parameter as the operand with the higher decimal precision.

If a binary numeric operator is applied to an integer operand and a real or complex operand, the result of the operation has the same kind type parameter as the real or complex operand, respectively.

If both operands are of type *default* real or *default* complex, the result also is of type default real or default complex, respectively.

Shape: If a binary numeric operator is applied to two operands of the same shape, the result of the operation also has this shape. If the shapes of the operands are different, the result has the same shape as the array operand.

Type, kind type parameter, and interpretation of $a+b$, $a-b$, $a*b$, and a/b are presented in the following table:

Operand types: $i=$ integer, $r=$ real, $c=$ complex.
Operator $\otimes$ stands for $+$ or $-$ or $*$ or $/$.

Operands			Result of $a \otimes b$		
a	b	Kind	Type	Kind	Interpretation
i	i	$ki(a) = ki(b)$	i	$ki(a)$	$a \otimes b$
i	i	$ra(a) < ra(b)$	i	$ki(b)$	$\mathrm{INT}(a, ki(b)) \otimes b$
i	i	$ra(a) > ra(b)$	i	$ki(a)$	$a \otimes \mathrm{INT}(b, ki(a))$
i	r		r	$ki(b)$	$\mathrm{REAL}(a, ki(b)) \otimes b$
i	c		c	$ki(b)$	$\mathrm{CMPLX}(a, ki(b)) \otimes b$
r	i		r	$ki(a)$	$a \otimes \mathrm{REAL}(b, ki(a))$
r	r	$ki(a) = ki(b)$	r	$ki(a)$	$a \otimes b$
r	r	$prec(a) < prec(b)$	r	$ki(b)$	$\mathrm{REAL}(a, ki(b)) \otimes b$
r	r	$prec(a) > prec(b)$	r	$ki(a)$	$a \otimes \mathrm{REAL}(b, ki(a))$
r	c	$ki(a) = ki(b)$	c	$ki(a)$	$\mathrm{CMPLX}(a, ki(a)) \otimes b$
r	c	$prec(a) < prec(b)$	c	$ki(b)$	$\mathrm{CMPLX}(a, ki(b)) \otimes b$
r	c	$prec(a) > prec(b)$	c	$ki(a)$	$\mathrm{CMPLX}(a, ki(a)) \otimes \mathrm{CMPLX}(b, ki(a))$
c	i		c	$ki(a)$	$a \otimes \mathrm{CMPLX}(b, ki(a))$
c	r	$ki(a) = ki(b)$	c	$ki(a)$	$a \otimes \mathrm{CMPLX}(b, ki(a))$
c	r	$prec(a) < prec(b)$	c	$ki(b)$	$\mathrm{CMPLX}(a, ki(b)) \otimes \mathrm{CMPLX}(b, ki(b))$
c	r	$prec(a) > prec(b)$	c	$ki(a)$	$a \otimes \mathrm{CMPLX}(b, ki(a))$
c	c	$ki(a) = ki(b)$	c	$ki(a)$	$a \otimes b$
c	c	$prec(a) < prec(b)$	c	$ki(b)$	$\mathrm{CMPLX}(a, ki(b)) \otimes b$
c	c	$prec(a) > prec(b)$	c	$ki(a)$	$a \otimes \mathrm{CMPLX}(b, ki(a))$

In this table, the type conversion functions INT, REAL, and CMPLX are the intrinsic functions defined in chapter 14; and the functions ki, $prec$, and ra have the same interpretation as the intrinsic functions KIND, PRECISION, and RANGE, respectively.

Type, kind type parameter, and interpretation of a ** b correspond to the last table with some exceptions:

Operand types: i = integer, r = real, c = complex.

Operands			Result of a ** b		
a	b	Kind	Type	Kind	Interpretation
i	i	RANGE(a) > RANGE(b)	i	KIND(a)	a ** b
r	i		r	KIND(a)	a ** b
c	i		r	KIND(a)	a ** b
other	other	other	corresponding to the last table		

If the operands of the exponentiation are of type integer and if the exponent is negative, the interpretation of a ** b is the same as the interpretation of $1 / (a ** \text{ABS}(b))$; ABS has the same interpretation as the intrinsic function ABS. This equivalent numeric expression is subject to the rules of *integer division* (see below).

2 ** (-3)   is interpreted as   1 / (2 ** 3)   and evaluates to 0

Invalid operations: A numeric operation is prohibited during program execution whose mathematically result is invalid in F; for example: division by zero, zero raised to zero power, zero raised to a negative power, and a negative real value raised to a real power.

Integer division

If the result of a division of two integer operands is *mathematically* not a whole number, the mathematical result is converted to integer type, that is to say, it is truncated towards zero. Thus the noninteger part of the mathematical result is lost.

8 / 3   evaluates to 2
(-8) / 3   evaluates to −2
- 100/99   evaluates to −1

7.2   Relational Intrinsic Expressions

A **relational intrinsic expression** compares the results of two numeric expressions or of two character expressions. A relational intrinsic expression must not compare a numeric expression with a logical or character expression, a logical expression with any intrinsic expression, or a character expression with any noncharacter expression.

17+4 <= nn   klm == airline + i3   "Jean" /= "Joan"

A relational expression may appear only as an operand in a logical expression. It evaluates either to a scalar value (that is, the value *true* or the value *false*) or to an array value of type default logical.

Relational intrinsic operators for numeric or character data entities:

Operator	Operation: comparison	Use	Interpretation
<	Less Than	x < y	$x < y$
<=	Less Than Or Equal To	x <= y	$x \leq y$
==	Equal To	x == y	$x = y$
/=	Not Equal To	x /= y	$x \neq y$
>	Greater Than	x > y	$x > y$
>=	Greater Than Or Equal To	x >= y	$x \geq y$

7.2.1 Numeric Relational Intrinsic Expressions

A **numeric relational intrinsic expression** compares the results of two numeric intrinsic expressions. If one operand is of type complex and the other one of any numeric type, only the operators == and /= are allowed.

A scalar numeric relational intrinsic expression evaluates to the default logical value *true* if and only if the operands satisfy the relation specified by the operator; otherwise the expression evaluates to the default logical value *false*. If the operands are conformable arrays, the result of the expression is produced element-wise for corresponding array elements.

When the types or kind type parameters of the operands in the numeric relational expression

numeric_expression₁ relational_operator numeric_expression₂

are different, the type and/or the kind type parameter of the result of one operand is internally converted in accordance to the rules for the evaluation of the expression (**numeric_expression₁ + numeric_expression₂**).

17 <= 13   evaluates to *false*.

abs(x-2.3) < 1.0e-75 tests a real value.

7.2.2 Character Relational Intrinsic Expressions

A **character relational intrinsic expression** compares the results of two character expressions lexically according to the ASCII collating sequence. The two operands may have diffent character lengths.

A scalar character relational intrinsic expression evaluates to the default logical value *true* if and only if the operands satisfy the lexical relation specified by the operator; otherwise the expression evaluates to the default logical value *false*. If the operands are conformable arrays, the result of the expression is produced element-wise for corresponding array elements.

The operands are compared one character at a time in order, beginning with the first character of each operand. If the operands have different lengths, the shorter one is treated as though it were extended on the right with blank characters to the length of the longer operand.

The result of the character relational intrinsic expression

character_expression₁ relational_operator character_expression₂

is given by the first character c_1 of the value of **character_expression₁** that is different from the corresponding character c_2 of the value of **character_expression₂**. The **character_expression₁** is (lexically) *less than* the **character_expression₂** if and only if the character position of c_1 precedes the character position of c_2 in the ASCII collating sequence. There are similar rules for the other relational operators. Two character expressions of length zero are always lexically equal when being compared in such relational expressions.

Collating sequence

According to the ASCII collating sequence (see appendix A), digits are in numeric sequence, letters are in alphabetic sequence, and the relation between digits, letters, and blank character is:

> blank character < digits < upper-case letters < lower-case letters

`"books" < "text"`	is *true*.
`"Text" > "Texts"`	is *false*.

For the programming of lexical comparisons, the intrinsic functions CHAR and ICHAR may also be used. CHAR returns for a given integer value the corresponding ASCII character, and ICHAR returns for a given character the integer value which is the position of this character in the ASCII collating sequence.

7.3 Logical Intrinsic Expressions

Logical intrinsic expressions are used to express logical computations. A logical expression evaluates either to a scalar logical value (that is, the value *true* or the value *false*) or to a logical array value. A **logical intrinsic expression** consists of *logical operands* and *logical operators*.

```
phi .or. lambda .and. psi .or. (.not. alpha(b))
```

Logical operands are logical constants, logical variables, logical array constructors, high precedence logical defined subexpressions (see 7.6.1), logical function references, relational intrinsic expressions, and logical (intrinsic and defined) (sub)expressions enclosed in parentheses.

Simple logical expressions may consist only of one operand without an operator. More complicated logical expressions consist of one or more operands which are processed by *logical intrinsic operators*. They may include logical subexpressions enclosed in parentheses. These parentheses must appear in pairs.

Logical intrinsic operators are

Operator	Operation	Use	Interpretation
.NOT.	Logical Negation	.not. x	*true* if x is *false*
.AND.	Conjunction	x .and. y	*true* if both x and y are *true*
.OR.	Inclusive Disjunction	x .or. y	*true* if x and/or y are *true*
.NEQV.	Non-equivalence	x .neqv. y	*true* if x is *true* and y is *false* or if y is *true* and x is *false*
.EQV.	Equivalence	x .eqv. y	*true* if both x and y are *true* or if both x and y are *false*

The precise interpretations of the logical intrinsic operators are given in the following truth tables:

x	y	.NOT. y	x .AND. y	x .OR. y	x .EQV. y	x .NEQV. y
true	*true*	*false*	*true*	*true*	*true*	*false*
true	*false*	*true*	*false*	*true*	*false*	*true*
false	*true*	*false*	*false*	*true*	*false*	*true*
false	*false*	*true*	*false*	*false*	*true*	*false*

Interpretation rules: If a logical intrinsic expression consists of several operands and operators, the expression is interpreted in accordance with the following rules:

- A logical intrinsic expression is interpreted from the left to the right. That is, if there is an operand between two operators of the same precedence, the left operator is combined with the operand.

- A parenthesized subexpression is treated as a data entity.

- If an expression includes operators of different precedence, the precedence of the operators controls the order of the combination of operands and operators.

Precedence of operators:

Operator	Precedence
.NOT.	highest
.AND.	high
.OR.	low
.EQV. or .NEQV.	lowest

`.not.a .or. b.and.c`   is interpreted as   `(.not.a) .or. (b.and.c)`

Logical expressions must not include two or more consecutive logical operators. But there are two exceptions which are demonstrated in the next example.

`1 .or. m .and. 2+4*8 <= 40 .neqv. p`   is (grammatically) correct.

`.true. .or. .eqv. .true.`   is not valid.

`r .and. .not. s` is allowed and is interpreted as `r .and. (.not. s)`.

`r .or. .not. s` is allowed and is interpreted as `r .or. (.not. s)`.

Once the interpretation of a logical intrinsic expression has been fixed, not all operands need be evaluated if the result of the expression may be determined in another way.

`x > y .or. (23-x*6 > x*y)`

If `x` is actually greater than `y`, the parenthesized subexpression need not be evaluated.

Kind type parameter and shape

Kind type parameter, shape, and value of a logical intrinsic expression result from those of the operand(s) and from the interpretation of the expression.

Unary logical operator: If the operator .NOT. is applied to a logical operand, the kind type parameter of the result of the operation is the same as the kind type parameter of the operand. There is a similar rule for the shape of the operand and of the result.

Binary logical operator: If a logical operator is applied to two logical operands with the same kind type parameter, the kind type parameter of the result of the operation is the same as the kind type parameters of the operands.

A binary logical operator may be applied to two operands having different shapes (if one is a scalar) or to two logical operands with different kind type parameters.

Kind type parameter conversion: If a binary logical operator is applied to two logical operands with different kind type parameters, the kind type parameter of the result is processor-dependent.

Shape: If a binary logical operator is applied to two operands of the same shape, the result of the operation also has this shape. If the operands have different shapes, the result has the same shape as the array operand.

7.4 Character Intrinsic Expressions

A *character expression* evaluates to a character value, that is, a value of type character. The result is either a scalar value (a character string) or an array value. A **character intrinsic expression** consists of *character operands* and *character operators*.

```
a // "B" // (c(d) // "EFG")
```

Character operands are character constants, character variables, character array constructors, high precedence character defined subexpressions (see 7.6.1), character function references, and character (intrinsic or defined) (sub)expressions enclosed in parentheses.

Simple character expressions may consist only of one operand without an operator:

```
"Example"        "EXAMPLE"         "1"        t10        line(s)
```

More complicated character intrinsic expressions consist of two or more operands which are processed by *character intrinsic operators*. They may include character (sub)expressions enclosed in parentheses. These parentheses must appear in pairs.

There is only one **character intrinsic operator**:

Operator	Operation	Use	Interpretation
//	Concatenation	x // y	Concatenate x with y

The character operands may have different character lengths.

A character expression has a character length which is the character length of the result of the expression. The result of the concatenation of two scalar character operands *c1* and *c2* is a character string with a length which is the sum of the lengths of the operands, $\text{LEN}(c1) + \text{LEN}(c2)$. The first part of the resulting string consists of the characters of the first operand *c1* and the second part consists of the characters of the second operand *c2*.

```
"Kinder" // "Garten"        evaluates to the string        KinderGarten
```

An F processor may evaluate only as much of a character intrinsic expression as is needed at this point to determine the result.

```
character (len=2) :: a, b, c, cf
⋮
a = b // cf(c)
```

The character function cf need not be evaluated, because a has character length 2 and the first operand b defines already a on the left-hand side of the equals.

Interpretation rules: If a character intrinsic expression consists of several operands and operators, the expression is interpreted from the left to the right. That is, if there is an operand between two operators, the left operator is combined with the operand. A parenthesized subexpression is treated as a data entity; but in this case, parentheses have no effect on the result of the evaluation.

```
"BOOK" // "WORM" // "LOVER"      is interpreted as
("BOOK" // "WORM") // "LOVER"
```

The result of the evaluation is the character string BOOKWORMLOVER.

Character length and shape

Character length, shape, and value of a character intrinsic expression result from those of its operands and from the interpretation of the expression.

Character length: The character lengths of the operands may be different. The length of the result of the concatenation is equal to the sum of the lengths of the operands.

Shape: If the concatenation operator is applied to two operands of the same shape, the result of the concatenation also has this shape. If the shapes of the operands are different, one operand must be scalar and the result has the same shape as the array operand.

7.5   Defined Expressions

A defined expression may consist of operands of derived and intrinsic types and of *defined operators*, intrinsic operators, and *extended* intrinsic operators. Note that a defined expression may also be a numeric, logical, or character expression if it evaluates to a result of such an intrinsic type.

A **defined expression** and an intrinsic expression differ in that a defined expression has at least one operand of derived type, or in that a defined expression includes a defined operator or an extended intrinsic operator.

Operands are (as in the case of an intrinsic expression) constants, variables, array constructors, function references, and (sub)expressions enclosed in parentheses. Defined expressions also may include structure constructors as operands.

New *operators* may be defined for operands of any data types. And the interpretation of such a defined operator may be extended. In a similar way, the interpretation of an intrinsic operator may also be extended. The extension of operators is sometimes described as "overloading".

7.5.1 Defined Operators and Extended Operators

For the definition of a new unary or binary *defined operator*, an *operator interface block* and at least one *operator function* must be supplied. The operator function defines the operation to be performed by the application of the operator. And the operator interface block specifies for a given operator which operator functions are responsible for the interpretation of the operator. Similar measures are necessary for the extension of an intrinsic operator.

Unary and binary defined operators may be used in the same way as intrinsic operators. That means, a unary defined operator appears immediately before its operand and a binary defined operator appears between its two operands.

A unary defined operator has the highest operator precedence, and a binary defined operator has the lowest operator precedence. But an extended intrinsic operator has the same precedence as the intrinsic operator.

Operator function

An **operator function** is a "normal" module function with one or two dummy arguments. The dummy arguments of operator functions must be non-optional arguments representing data objects. They must be specified with the INTENT(IN) attribute.

The name of an operator function must be explicitly specified in a PUBLIC or PRIVATE statement within the specification part of the module.

Where an operator function is available, it may be referenced like any other function as an operand in an expression. And in a scoping unit where an operator interface block is available, this function may be invoked implicitly (i. e. automatically) when the associated operator is encountered during expression evaluation.

Operator interface block

An **operator interface block** is a special *generic interface block* for a (nonextended or an extended) defined operator or for an extended intrinsic operator. Such an interface block may appear only in the specification part of a private module.

INTERFACE OPERATOR (operator)   ⟵ INTERFACE statement

  MODULE PROCEDURE ...   ⟵ MODULE PROCEDURE stmt(s).
  ⋮

END INTERFACE   ⟵ END INTERFACE statement

where **operator** is the operator to be defined or extended. It is either an intrinsic operator or it is a defined operator written as **.op.**, where **op** is a sequence of 1 to maximal 31 letters. Such a designator of a defined operator must be different from the logical literal constants .TRUE. and .FALSE..

The *specific name* of an operator function must appear only once in all operator interface blocks being available for the same operator in a scoping unit. The generic specifier OPERATOR(...) must be explicitly specified in a PUBLIC or PRIVATE statement in the specification part of the module with the interface block.

The MODULE PROCEDURE statement has the following form:

MODULE PROCEDURE operator_function [, operator_function]...

where each **operator_function** is the name of a module function.

Note that two or more interface blocks for the same operator being available in a scoping unit are interpreted as a single generic interface block.

Implicit reference

An operator in an expression is interpreted in a scoping unit as a defined operator or as an extended intrinsic operator (that is, as an implicit reference to an operator function),

- If an operator interface block for this operator is accessible in the scoping unit using the operator;

- If the operator function specified in the operator interface block can be referenced in the scoping unit;

- If the operand(s) agree with the corresponding dummy argument(s) of the operator function with regard to data type, (if applicable) kind type parameter, and (if applicable) character length; and

- If the operand(s) and the corresponding dummy argument(s) agree with regard to rank and, if applicable, shape.

If this is true for a unary operator, the operand is used as the actual argument in the implicit reference to the operator function. And in the case of a binary operator, the left operand is used as the first actual argument in the implicit reference to the operator function and the right operand is used as the second actual argument.

7.5.1.1   Nonextended Defined Operator

The form of an **operator interface block** for a nonextended defined operator **.op.** is:

INTERFACE OPERATOR (.op.)	←— INTERFACE statement
MODULE PROCEDURE operator_function	←— MODULE PROCEDURE st.
END INTERFACE	←— END INTERFACE stmt.

```
module def_op
  private :: width
  public  :: operator(.dist.)
  interface operator (.dist.)
    module procedure width
  end interface
contains
  function width (a, e) result (distance)
    real, dimension (2), intent(in) :: a, e
    real :: distance
    distance = sqrt( abs( (a(1) - e(1))**2 - (a(2) - e(2))**2))
  end function width
end module def_op

program flight
use def_op
:
if ((la .dist. hannover) > (la .dist. frankfurt)) then
  call departure(frankfurt)
endif
:
end program flight
```

The defined operator .dist. is a binary operator whose interpretation is given by the operator function width. Since the operator function has the PRIVATE attribute, it could not be explicitly referenced outside the module.

An operator may be defined for operands of intrinsic types, but it may also be defined for operands of derived types.

7.5.1.2 Extended Defined Operator

A defined operator may be extended by specifying more than one operator function in the operator interface block or by accessing two or more interface blocks for the operator in the scoping unit.

All these functions specified in the operator interface block for a particular operator must be valid operator functions for this operator. They must have the same number of dummy arguments. And they must satisfy similar unambiguity rules as those specific subprograms which have a common generic name: within a scoping unit, any two operator functions for the same operator must have a dummy argument at the same position such that these two arguments have a different data type, (if applicable) kind type parameter, or rank.

When such an extended defined operator is encountered, the characteristics of the operands determine which of these operator functions will actually be invoked and executed.

7.5.1.3 Extended Intrinsic Operator

If the operator specified in the INTERFACE statement of an operator interface block is an intrinsic operator, this defines an extension of the intrinsic operator. An operator function for such an extended intrinsic operator may extend the operator only for those data types of its operands that do not belong to the data types for which this operator is predefined.

The number of arguments of an operator function for an extended intrinsic operator must agree with the number of operands needed for the intrinsic operator. That is, a unary intrinsic operator may be extended only unary, and a binary intrinsic operator may be extended only binary.

There are similar unambiguity rules as for extended defined operators. In the case of an extended intrinsic operator, the unambiguity rules are applied both to the operator functions which are specified in the interface block(s) for this extended intrinsic operator and to an imaginary collection of additional operator functions defining the predefined meaning of the intrinsic operator. For any valid combination of types, (if applicable) kind type parameters, and ranks of the operands, a corresponding operator function could be thought.

When the extended intrinsic operator is encountered, the characteristics of the operands determine which of these operator functions will actually be invoked and executed.

7.6   Common Rules for Expressions

7.6.1   Precedence of Operators

If there are numeric, logical, relational, or character intrinsic operators, and/or defined operators in the same expression, the combination of operands with operators is determined by the *precedence of operators.*

Expression	Operator	Operation	Precedence Level
High Precedence Defined Expressions	*Unary* Def. Op.	Operator Function	1. highest
Numeric Intrinsic Expressions	**	Exponentiation	2.
	*	Multiplication	3.
	/	Division	3.
	+	Identity	4.
	−	Numeric Negation	4.
	+	Addition	5.
	−	Subtraction	5.
Character Intrinsic Expressions	//	Concatenation	6.
Relational Intrinsic Expressions	==	Equal To	7.
	>=	Greater Than Or Equal To	7.
	>	Greater Than	7.
	<=	Less Than Or Equal To	7.
	<	Less Than	7.
	/=	Not Equal To	7.
Logical Intrinsic Expressions	.NOT.	Logical Negation	8.
	.AND.	Conjunction	9.
	.OR.	Inclusive Disjunction	10.
	.EQV.	Equivalence	11.
	.NEQV.	Non-equivalence	11.
Low Precedence Defined Expressions	*Binary* Def. Op.	Operator Function	12. lowest

If an intrinsic operator is extended, this extended operator has the same precedence as the intrinsic operator, even when the operator is used with its extended interpretation. If operators and operands are to be combined without regard to the rules of operator precedence, parentheses may be used.

```
r .and. .not. s        5 + .fahrtocel. 13        - .celtofahr. n
```

7.6.2   Interpretation of Expressions

Operators are combined with operands according to the following interpretation rules in this order:

1. A parenthesized (sub)expression is treated as a data entity.

2. The precedence of the operators determines how operands are combined with operators in expressions with several operators.

3. Successive exponentiations are combined from the right to the left.

4. Successive multiplications and/or divisions are combined from the left to the right.

5. Successive additions and/or subtractions are combined from the left to the right.

6. Successive concatenations are combined from the left to the right.

7. Successive logical conjunctions (logical products) are combined from the left to the right.

8. Successive logical inclusive disjunctions (logical sums) are combined from the left to the right.

9. Successive logical equivalences and/or non-equivalences are combined from the left to the right.

10. Successive binary defined operators are combined from the left to the right.

```
a + b ** 2 < 7 .or. 1 .and. c == "letter" // "head"
```

is interpreted as

```
((a + (b ** 2)) < 7) .or. (1 .and. (c == ("letter" // "head")))
```

Once the F processor has established the interpretation of an expression, it may evaluate an equivalent expression (see page 7-2 "Interpretation").

```
3 / 4 * 8      evaluates to   0
3 * 8 / 4      evaluates to   6
```

The last two expressions are not equivalent (according to F rules). The result of the evaluation depends on the order of the evaluation.

7.6.3   Evaluation of Expressions

Operands

A scalar variable must not be used as an operand in an expression until it is defined with a valid value. Accordingly, all elements of an array must be defined when this array is used as an operand.

When a variable of derived type is used as an operand, all components of the variable must be defined. All characters of a character operand must be defined when it is used as an operand.

If an operand is a pointer or a pointer enclosed in parentheses, the associated target is used in the expression. Type, kind type parameter (if applicable), length (if applicable), shape, and value of the operand are those of the associated target.

Elemental operations

Elemental operations on array operands are performed element-wise as a set of scalar operations which may be executed in any order. The operand(s) being involved in such a scalar-like operation and the result are corresponding array elements, that is to say, elements in the same array element position within their respective parent array. When a binary elemental operator has a scalar operand and an array operand (with at least one array element), the scalar operand is treated as if it were expanded into an array whose shape is that of the array operand and whose array element values are equal to that of the scalar operand.

Whether or not an extended intrinsic operator is interpreted as an elemental operation may depend on the encountered operand(s) during program execution.

Side effects

Certain *side effects* are not allowed. The evaluation of a function must not change any data entity appearing in the expression with the function reference except the function result.

```
a(i) = f(i)
y = g(x) + x
```

If the function reference f(i) modifies the actual argument i, or if the function reference g(x) modifies the actual argument x, the respective assignment statement is not valid.

The evaluation of an actual argument must not affect the type of the expression with the function reference and the type of the expression must not affect the evaluation of the actual argument; exception: generic functions. Note that the type of the result of a generic function may depend on the types of the arguments in the function reference.

If array elements or array sections appear as operands, subscript expressions are to be evaluated. The evaluation of a subscript expression does neither affect nor is affected by the type of the expression with the array element or array section. There is a similar rule for the evaluation of character substring expressions. The appearance of an array constructor in an expression may require the evaluation of the parameters of an included implied-DO. The type of the expression with this array constructor does neither affect nor is affected by the evaluation of the implied-DO parameters.

7

Zero-size arrays and zero-length character strings

An F processor needs to evaluate neither a subscript expression of a zero-size array nor a character substring expression of a zero-length substring if such an array or substring, respectively, appears as an operand in an expression.

Evaluation of single operands in expressions with several operands

Not all operands in an expression need be evaluated and an operand need not be completely evaluated if the result of the expression can be determined in a different way. Such a situation may arise, for example, during the evaluation of a logical expression or if an operand is a zero-size array or a zero-length character string.

```
x > y .or. a+b*c <= 17 .or. l(z)
```

If the F processor knows that x is actually greater than y, the rest of the logical expression on the right-hand side of the first .or. need not be evaluated.

```
x + w(z)
```

If x is a zero-size array, w(z) need not be evaluated.

If a statement includes an expression with a function reference which need not be evaluated for the execution of the statement, then all those data objects become undefined after the evaluation of the expression which might have become defined by the evaluation of the function reference.

```
xx + ww(z1)
x > y .or. a+b*c <= 17 .or. 11(z2)
```

Suppose, xx is a zero-size array and x > y is true: If the function ww would define the actual argument z1 or if function 11 would define the actual argument z2, then z1 and z2, respectively, are undefined after the evaluation of the respective expression.

```
character (len=2) :: a, b, c, f
a = b // f(c)
```

The character function f need not be evaluated because the value of b determines the value for the left-hand side a completely.

Order of evaluation of function references

An F processor may evaluate function references appearing in one statement in any order. And the result of each function reference must be independent of the order of the evaluations of the functions; exception: function references in the actual argument list of function references.

```
y = f( 1(z))
```

The function 1 is evaluated first and then (possibly) the function f.

7.7   Special Expressions

For the specification of constant values, sometimes not only (literal or named) constants but even more complicated *constant expressions* may be written. For the initialization of variables and for the definition of named constants within type declaration statements and other specification statements, *initialization expressions* may be written. These two classes of special expressions can be evaluated already during program compilation. *Specification expressions* form a third class of special expressions which may be used to specify certain properties of data objects in the specification part of subprograms. These last forms of special expressions can be evaluated during program execution on invocation of the subprogram.

7.7.1   Constant Expressions

A **constant expression** is an expression in which each operator is intrinsic and each of its operands is one of the following constant data entities:

- A constant or a subobject of a named constant,

- An array constructor where each element and the loop parameters of each contained implied-DO are expressions whose operands are constant expressions,

- An implied-DO variable within an array constructor where the loop parameters of the implied-DO are constant expressions,

- A structure constructor where each component is a constant expression,

- An elemental intrinsic function reference where each actual argument is a constant expression,

- A transformational intrinsic function reference where each actual argument is a constant expression,

- A reference to one of the intrinsic inquiry functions LBOUND, SHAPE, SIZE, UBOUND, LEN, BIT_SIZE, KIND, DIGITS, EPSILON, HUGE, MAXEXPONENT, MINEXPONENT, PRECISION, RADIX, RANGE, and TINY, where each actual argument is either a constant expression or a variable whose kind type parameters, character lengths, and bounds inquired about are not assumed, are not defined by an expression that is not a constant expression, and are not definable by an ALLOCATE statement or a pointer assignment statement, or

- A constant expression enclosed in parentheses,

where each subscript, section subscript, substring starting character position, and substring ending character position is a constant expression. Any other named or unnamed variables, any non-intrinsic function reference, and any defined operator are prohibited within constant expressions.

Depending on the data type of the expression, there are **integer constant expressions**, **numeric constant expressions**, **logical constant expressions**, and **character constant expressions**.

```
2                                        ← integer constant expression
range(i) / 10                            ← integer constant expression
count( (/ .true., .false., .false. /) )  ← integer constant expression
- 23.5 + (4 * 47) ** 2                   ← real constant expression
1.0 - abs(-0.4)                          ← real constant expression
.true.                                   ← logical constant expression
sqrt(63.3) - 8.2 < 0.0                   ← logical constant expression
"timetable"                              ← character constant expr.
"time" // " sheet"                       ← character constant expr.
```

7.7.2   Initialization Expressions

An **initialization expression** is a constant expression in which the exponentiation is permitted only with an integer power and each of its operands is one of the following:

- A constant or a subobject of a named constant,

- An array constructor where each element and the parameters of its contained implied-DOs are expressions whose operands are initialization expressions,

- An implied-DO variable within an array constructor where the loop parameters of the implied-DO are initialization expressions,

- A structure constructor where each component is an initialization expression,

- A reference to an elemental intrinsic function of type integer or character where each actual argument is an initialization expression of type integer or character,

- A reference to one of the transformational intrinsic functions REPEAT, RESHAPE, SELECTED_INT_KIND, SELECTED_REAL_KIND, and TRIM, where each actual argument is an initialization expression,

- A reference to one of the intrinsic inquiry functions LBOUND, SHAPE, SIZE, UBOUND, LEN, BIT_SIZE, KIND, DIGITS, EPSILON, HUGE, MAXEXPONENT, MINEXPONENT, PRECISION, RADIX, RANGE, and TINY, where each actual argument is either an initialization expression or a variable whose properties inquired about are not assumed, are not defined by an expression that is not an initialization expression, and are not definable by an ALLOCATE statement or a pointer assignment statement, or

- An initialization expression enclosed in parentheses,

where each subscript, section subscript, substring starting character position, and substring ending character position is an initialization expression. Any other named or unnamed variables, any non-intrinsic function references, and any defined operators are prohibited within initialization expressions.

```
2                                          ←— integer initialization expr.
range(i) / 10                              ←— integer initialization expr.
count( (/ .true., .false., .false. /) )←— no initialization expression,
                                              prohibited transformational
                                              intrinsic function
- 23.5 + (4 * 47) ** 2                     ←— real initialization expression
```

```
1.0 - abs(-0.4)                    ←— no initialization expression,
                                        real actual argument
.true.                             ←— logical initialization expr.
sqrt(63.3) - 8.2 < 0.0             ←— logical initialization expr.
"timetable"                        ←— character initialization expr.
"time" // " sheet"                 ←— character initialization expr.
```

If an initialization expression includes a reference to an inquiry function for a kind type parameter, a character length, or an array bound of an object specified in the same specification part, the kind type parameter, length, or bound must be specified lexically earlier in the specification part. This prior specification may appear to the left of the inquiry function in the same statement.

7.7.3 Specification Expressions

In type declaration statements and in other specification statements within the specification part of a subprogram, *specification expressions* may be used to specify array bounds and character lengths. These specification expressions may contain not only "normal" operands but also operands which are composed of other data entities, for instance array constructors, structure constructors, subprogramm references, and subexpressions in parentheses. The constituent parts of these compound operands may be written as *restricted expressions*. Specification expressions are special forms of restricted expressions. Both are subject to particular restrictions because they are evaluated during program execution after invocation and before execution of the subprogram.

A **restricted expression** is an expression in which each operator is intrinsic and each of its operands is one of the following:

- A constant or a subobject of a named constant,

- A variable which is a dummy argument that has neither the OPTIONAL nor the INTENT(OUT) attribute, or a variable that is a subobject of such a dummy argument,

- A variable that is made available by USE association or host association or a variable that is a subobject of such a variable,

- An array constructor where each element and the parameters of its contained implied-DOs are expressions whose operands are restricted expressions,

- An implied-DO variable within an array constructor where the loop parameters of the implied-DO are restricted expressions,

- A structure constructor where each component is a restricted expression,

- A reference to an elemental intrinsic function of type integer or character where each actual argument is an restricted expression of type integer or character,

- A reference to one of the transformational intrinsic functions REPEAT, RESHAPE, SELECTED_INT_KIND, SELECTED_REAL_KIND, and TRIM, where each actual argument is a restricted expression,

- A reference to one of the intrinsic inquiry functions LBOUND, SHAPE, SIZE, UBOUND, LEN, BIT_SIZE, KIND, DIGITS, EPSILON, HUGE, MAXEXPONENT, MINEXPONENT, PRECISION, RADIX, RANGE, and TINY, where each actual argument is either a restricted expression or a variable whose properties inquired about are not defined by an expression that is not a restricted expression, and are not definable by an ALLOCATE statement or a pointer assignment statement, or

- A restricted expression enclosed in parentheses,

where each array subscript, section subscript, substring starting character position, and substring ending character position is a restricted expression. Any other named or unnamed variables, any references to another class of functions, and any defined operators are prohibited within restricted expressions.

A **specification expression** is a scalar restricted expression of type integer.

```
len(line) - 2        ! line is a character variable
kind(x) - kind(y)    ! x and y are real variables
ubound(f, 2)         ! f is a 2-dimensional local array
```

Type, (if applicable) kind type parameter, and (if applicable) character length of a variable in a specification expression must be explicitly declared lexically earlier in this scoping unit, or the declaration must be available by USE association or host association.

If a specification expression includes a reference to a function inquiring about a kind type parameter, a character length, or an array bound of an entity in the same specification part, this kind type parameter, this character length, and this array bound, respectively, must be specified lexically earlier in this specification part. If a specification expression includes a reference to the value of an element of an array specified in the same specification part, the array bounds must be specified lexically earlier in this specification part. The prior specification may be on the left of the inquiry function reference in the same statement.

8 ASSIGNMENTS

Assignments are executable statements. There are *intrinsic assignment statements*, *defined assignment statements*, *pointer assignment statements*, and *masked array assignment statements*.

During execution of an intrinsic assignment statement, a variable becomes defined (with a value) or redefined (with a new value). During execution of a defined assignment statement, a variable may become defined or redefined; other actions are also allowed. During execution of a pointer assignment statement, a pointer may become associated with a target such that afterwards the pointer points to that target. During execution of a masked array assignment (WHERE construct), one or more array assignment statements are evaluated element-for-element for those array element positions that are selected by a given logical array expression.

8

8.1 Intrinsic Assignment Statements

Intrinsic assignment statements cannot be identified by characteristic keywords. The form of an intrinsic assignment statement is:

variable = expression

The **expression** is evaluated first. Then its result is assigned (possibly after certain conversions) to the **variable**. The two parts of the assignment statement are called the *left-hand side* and the *right-hand side*.

The intrinsic *numeric assignment statement*, the *logical assignment statement*, the *character assignment statement*, and the *assignment statement for derived types* are predefined; that is, the interpretation of the assignment symbol is given by the language. For these intrinsic assignment statements, the left-hand side and the right-hand side must be conformable and both sides must be of numeric type, of type logical, of type character, or of the same derived type.

Intrinsic array assignment statement: If the left-hand side of an assignment statement is an array, the statement is an **array assignment statement**. The left-hand side may be an array section but must not be an array section with a vector subscript with two or more array elements having the same value. If the left-hand side is a zero-size array, no value is assigned to the left-hand side.

If the left-hand side is an array and the right-hand side is scalar, all elements of the left-hand side are defined with the same scalar value of the right-hand side. If the right-hand side is an array value, the left-hand side must also be an array and both sides must have the same shape.

The left-hand side of an array assignment statement is defined element-wise for corresponding array elements; that is, the first element of the left-hand side is defined with the value of the first element of the right-hand side, the second element of the left-hand side is defined with the value of the second element of the right-hand side, and so on. The order of these element-wise assignments is processor-dependent.

```
real, dimension (10), save :: x = (/ (i, i=1,10) /)
x(1:10) = x(10:1:-1)
```

After execution of the array assignment statement, array x is defined with the values 10, 9, 8, ... , 2, 1.

Interpretation: The assignment (i. e., the storing of the value of the right-hand side at the address of the left-hand side) is not done until all evaluations of the right-hand side are completed and all address computations for subobjects (for instance array elements or character substrings) of the left-hand side are completed. Note that the same data objects may appear on both sides and that even the same subobjects may be affected on both sides. But side effects are not allowed; that is, the evaluation of expressions included in the left-hand side must neither affect nor be affected by the evaluation of the right-hand side.

The equals in an intrinsic assignment statement does not have the normal mathematical interpretation.

```
n = n + 1
```

is not a contradiction in the F language. The interpretation of this statement is as follows: increase the (current) value of the variable n by 1; then store the result of this addition in the address of the variable n. Thus n is redefined with a new value and its former value is lost.

Pointers: If the left-hand side is a pointer, this pointer must be pointer associated with a definable target at the time of the execution of the assignment statement. Type, kind type parameter (if applicable), character length (if applicable), and shape of the target must conform with those of the right-hand side. And the result of the evaluation of the right-hand side is assigned to the target the left-hand side is pointing at. If the right-hand side consists only of a pointer, the associated target is referenced.

8.1.1   Numeric Assignment Statement

If both the left-hand side and the right-hand side are of numeric type, it is an **numeric intrinsic assignment statement**.

numeric_variable = numeric_expression

The left-hand side and the right-hand side may be of different numeric types and may have different kind type parameters.

```
real :: x, var, f, y
⋮
x = 17 + var * 2 - f(y)
```

When a numeric assignment statement is executed, the following steps are performed in sequence:

1. Evaluation of the numeric expression on the right-hand side and all expressions used to determine the variable on the left-hand side.

2. Conversion of the result of the right-hand side according to the data type and/or the kind type parameter of the left-hand side if types and/or the kind type parameters of both sides are different.

3. Assignment (i. e., storing) of the (possibly converted) result of the right-hand side to the numeric variable (that is, in the address) of the left-hand side.

```
integer, parameter :: long  = selected_real_kind(14, 200)
integer, parameter :: short = selected_real_kind(7, 200)
real (kind= long)  :: dup
real (kind= short) :: xray, pi, r
⋮
dup = 4.7e3 * xray - 2.5 * r * pi
r   = 3.1415 - dup
```

These numeric assignment statements are interpreted as

```
dup = real(4.7e3 * xray - 2.5 * r * pi, kind(dup))
r   = real(3.1415 - dup, kind(r))
```

with REAL and KIND being intrinsic functions.

If a right-hand side with less decimal precision is assigned to a left-hand side with more decimal precision , only the internal representation of the result of the right-hand side is converted. Note that the precision of the internal representation of the right-hand side is not improved during the assignment operation.

8.1.2 Logical Assignment Statement

If both the left-hand side and the right-hand side are of type logical, it is a **logical intrinsic assignment statement**.

logical_variable = logical_expression

The left-hand side and the right-hand side may have different kind type parameters.

```
logical :: a, b, c, d
⋮
a = (b .or. c) .and. d
```

When a logical assignment statement is executed, the following steps are performed in sequence:

1. Evaluation of the logical expression on the right-hand side and all expressions used to determine the variable on the left-hand side.

2. Conversion of the result of the right-hand side according to the kind type parameter of the left-hand side if the kind type parameters of both sides are different.

3. Assignment (i. e., storing) of the (possibly converted) result of the right-hand side to the logical variable (that is, in the address) of the left-hand side.

8.1.3   Character Assignment Statement

If both the left-hand side and the right-hand side are of type character, it is a **character intrinsic assignment statement**.

character_variable = character_expression

The left-hand side and the right-hand side may have different character lengths. If the left-hand side has character length zero, no value is assigned.

```
character (len=12), save :: file = "result"
file = file(1:6) // ".f90"
```

The left-hand side and the right-hand side may overlap; that is, character positions may appear on the left-hand side which are used on the right-hand side.

When a character assignment statement is executed, the following steps are performed in sequence:

1. Evaluation of the character expression on the right-hand side and all expressions used to determine the variable on the left-hand side.

2. Conversion of the result of the right-hand side if both side have different character lengths: if the length of the **character_variable** is greater than the length of the **character_expression**, the result of the character expression is filled with trailing blank characters until it has the same length as the left-hand side; if the length of the **character_variable** is less than the

length of the **character_expression**, the result of the character expression is truncated from the right until it has the same length as the left-hand side.

3. Assignment (i. e., storing) of the (possibly converted) result of the right-hand side to the character variable (that is, in the address) of the left-hand side.

The value of the character expression of the right-hand side need be defined only for character positions which are used to define the left-hand side.

```
character (len=2)       :: a
character (len=4), save :: b = "ok"
a = b
```

For this character assignment statement, only positions b(1:1) and b(2:2) need be defined. It is not necessary that b(3:4) is defined.

8.1.4   Assignment Statement for Derived Types

When both the left-hand side and the right-hand side are of the same derived type, the statement is an **intrinsic assignment statement for derived types** if no assignment interface block and no assignment subroutine are available for this derived type in the scoping unit with this assignment statement.

derived_type_variable = defined_derived_type_expression

```
type, public :: employee
  character (len=15) :: name
  character (len=10) :: first_name
  integer :: number
  logical :: married
  real    :: salary
end type employee
```

Where this derived type is available, one can write the following statements:

```
type (employee), dimension (1000) :: region_north
region_north(78) = employee("Smith", "Peter", 78, .true., 840.64)
```

When an assignment statement for derived types is executed, the following steps are performed in sequence:

1. Evaluation of the defined expression on the right-hand side and all expressions used to determine the variable on the left-hand side.

2. Assignment (i. e., storing) of the result of the right-hand side to the variable (that is, in the address) of the left-hand side.

An intrinsic assignment statement for derived types is interpreted such as if the assignment is executed component-wise (for corresponding components). A nonpointer component is treated according to the rules of intrinsic assignments. And a component with POINTER attribute is treated according to the rules for *pointer assignment statements*.

The last assignment statement is interpreted as the following set of component-wise assignments:

```
region_north(78)%name       = "Smith"   ! character assignment
region_north78)%first_name  = "Peter"   ! character assignment
region_north(78)%number     = 78        ! numeric assignment
region_north(78)%married     = .true.    ! logical assignment
region_north(78)%salary     = 840.64    ! numeric assignment
```

8.2 Defined Assignment Statements

The defined assignment statement has the same form as an intrinsic assignment statement:

variable = expression

If the left-hand side and the right-hand side are *not both* of numeric type, *not both* of type logical, *not both* of type character, and *not both* of the same derived type, or if both sides are of the same derived type but an *assignment interface block* and an *assignment subroutine* for this type (see below) are available, then the statement is a **defined assignment statement**.

In this case, the equals does not have the "normal" intrinsic interpretation but its interpretation (as an assignment operator for this left-hand side and this right-hand side) is determined by an *assignment subroutine* and an *assignment interface block*, which must be available in the scoping unit with this assignment statement. The assignment subroutine specifies which operations are to be performed for the execution of the defined assignment statement. For example, the assignment subroutine may specify how the result of the right-hand side should be treated before it is assigned (or not) to the left-hand side. And the assignment interface block specifies the assignment subroutine(s) belonging to the defined assignment statement.

The interpretation of *intrinsic* assignment statements cannot be overridden by a defined assignment statement. Defined assignments are extensions of the intrinsic interpretation of the assignment symbol.

Assignment subroutine

An **assignment subroutine** is a "normal" module subroutine with exactly two dummy arguments.

The two dummy arguments must be nonoptional dummy arguments representing data objects. The first dummy argument must have the INTENT(OUT) or INTENT(INOUT) attribute and the second dummy argument must have the INTENT(IN) attribute.

The name of an assignment subroutine must be explicitly specified in a PUBLIC or PRIVATE statement within the specification part of the module.

Where an assignment subroutine is available, it may be referenced like any other subroutine by the execution of a CALL statement. And in a scoping unit where an appropriate assignment interface block for a defined assignment statement is available, the assignment subroutine may be invoked implicitly (i. e. automatically) during execution of the assignment statement.

Assignment interface block

An **assignment interface block** is a special *generic interface block* for a defined assignment. Such an interface block may appear only in the specification part of a private module.

INTERFACE ASSIGNMENT (=)	← INTERFACE statement
MODULE PROCEDURE ...	← MODULE PROCEDURE statement(s)
⋮	
END INTERFACE	← END INTERFACE statement

The *specific name* of an assignment subroutine must appear only once in all assignment interface blocks being available in a scoping unit. The generic specifier ASSIGNMENT(=) must be explicitly specified in a PUBLIC or PRIVATE statement in the specification part of the module with the interface block.

The MODULE PROCEDURE statement has the following form:

MODULE PROCEDURE assign_subr [, assign_subr]...

where each **assign_subr** is the name of a module subroutine.

Note that two or more assignment interface blocks being available in a scoping unit are interpreted as a single generic interface block.

Implicit reference

An assignment statement is interpreted as a defined assignment statement (that is, as an implicit reference to an assignment subroutine)

- If an assignment interface block is accessible in the scoping unit with the assignment statement;

- If the assignment subroutine is specified in the assignment interface block and can be referenced in the scoping unit;

- If the left-hand side and the right-hand side of the assignment statement agree with the corresponding dummy arguments of the assignment subroutine with regard to data type, (if applicable) kind type parameter, and (if applicable) character length; and

- If the left-hand side and the right-hand side and the corresponding first and second dummy argument agree with regard to rank and, if applicable, shape.

If this is true, the left-hand side is used as the first actual argument and the result of the evaluation of the right-hand side is used as the second actual argument for the implicit reference to the assignment subroutine.

8.2.1 Nonextended Defined Assignment

The form of an **assignment interface block** for a nonextended defined assignment is:

INTERFACE ASSIGNMENT (=)   ←— INTERFACE statement
   MODULE PROCEDURE assign_subr   ←— MODULE PROCEDURE stmt.
END INTERFACE   ←— END INTERFACE statement

```
module def_assign
  public :: assignment(=), integer_to_logical
  interface assignment (=)
    module procedure integer_to_logical
  end interface
contains
  subroutine integer_to_logical (left, right)
    logical, intent(out) :: left
    integer, intent(in)  :: right
    if (right /= right/2*2) then
      left = .false.
```

```
    else
      left = .true.
    endif
  end subroutine integer_to_logical
end module def_assign

program even
use def_assign
logical :: ls
integer :: julian
:

ls = julian - 15    ! defined assignment
:

end program even
```

The statement `ls = ...` is an assignment statement with a logical left-hand side
and with an integer right-hand side. The interpretation of this assignment operation is given by the assignment subroutine `integer_to_logical`.

8.2.2   Extended Defined Assignment

A defined assignment statement may be extended by specifying more than one
assignment subroutine in the assignment interface block or by accessing two or
more assignment interface blocks in a scoping unit.

All these subroutines specified in the assignment interface block must be valid
assignment subroutines. And they must satisfy similar unambiguity rules as
specific subprograms having a common generic name: Within a scoping unit, any
two assignment subroutines must have a dummy argument at the same position
such that these two arguments have a different data type, (if applicable) kind
type parameter, or rank.

When such an extended assignment is encountered, the characteristics of the
left-hand side and of the right-hand side determine which of the assignment
subroutines will be actually invoked and executed.

8.3   Pointer Assignment Statement

A **pointer assignment statement** may be used to associate a pointer with a
(new) target. Afterwards the pointer points at this new target and not at the
former target. The form of a pointer assignment statement is

pointer => target

The **target** may be one of the following:

– a variable having the TARGET attribute,
– a pointer, that is, a variable having the POINTER attribute, or
– an expression which evaluates to a pointer.

The **pointer** is called the *left-hand side* and the **target** is called the *right-hand side* of the pointer assignment statement.

The right-hand side must have either the POINTER attribute or the TARGET attribute. It may be a subobject of a variable with the TARGET attribute. But it must not be an array section with a vector subscript. If it is a structure component of a derived type variable without the TARGET attribute, the corresponding type component must have the POINTER attribute.

The right-hand side must have the same type, the same kind type parameter (if applicable), the same character length (if applicable), and the same rank as the pointer on the left-hand side.

If the right-hand side is *not* a pointer, the pointer on the left-hand side becomes associated with the target on the right-hand side; that is, the pointer points at the target on the right-hand side. In this case, the right-hand side may be even an allocatable array (with TARGET attribute).

If the right-hand side is a pointer, the pointer on the left-hand side receives the same pointer association status as the pointer on the right-hand side. If the right-hand side is a pointer which is currently associated, the pointer on the left-hand side becomes associated with the same target as the pointer on the right-hand side. If the right-hand side is a pointer which is currently not associated with a target, the pointer on the left-hand side also becomes "disassociated". And if the right-hand side is a pointer with undefined association status, the left-hand side also receives the pointer association status "undefined".

```
module ass
public :: f
type, public :: struct
  integer           :: i
  real, pointer     :: comp
  character (len=78) :: text
end type struct
contains
  function f (da1, da2) result (ff)
    real, intent(in) :: da1, da2
    real, pointer    :: ff
    ⋮
  end function f
end module ass
```

Where these module entities are available, one can declare

```
real, pointer :: line (:), short
real, target, dimension (45, 10) :: page
type (struct) :: structure
```

and can then write the following assignment statements:

```
line  => page(7, 1:10)          ! pointer => no pointer
short => structure%comp         ! pointer => pointer
structure%comp => f(aa1, aa2)   ! pointer => function result
```

8.4 Masked Array Assignments

Sometimes, array assignment statements must be executed only for selected elements, for example:

```
real, dimension (100) :: denominator, reci
read (unit=14) denominator
reci = 1/denominator
```

Suppose, the used F processor allows the execution of the above array assignment statement only for those array elements for which the denominator(i) is nonzero. If the array expression and the array assignment is to be preserved, a WHERE construct may be used containing an appropriate logical mask to avoid the illegal element-wise assignment operations. Such a statement is called a *masked array assignment.*

8.4.1 WHERE Construct

The WHERE construct statement, the END WHERE statement, and an optional ELSEWHERE statement may be used to control the execution of one or two sequences of array assignment statements by one logical array mask. Such a sequence of array assignment statements is called an **assignment block**.

The WHERE construct statement is the first statement of a WHERE construct. The END WHERE statement is the last statement of a WHERE construct. And one ELSEWHERE statement may be used if two assignment blocks are to be executed such that each of these assignment blocks is executed for another selection of array element positions.

The most simple form of a WHERE construct begins with a WHERE construct statement, ends with an END WHERE statement, and includes an assignment block with a sequence of intrinsic array assignment statements.

```
WHERE (mask_expression)
    array_variable₁ = expression₁
    array_variable₂ = expression₂
    ⋮
END WHERE
```

The where-block may be empty. The **mask_expression** is a logical array expression. The mask expression and the left-hand side of each array assignment statement within the body of the WHERE construct must have the same shape.

During execution of this WHERE construct, the mask expression is evaluated first; its result is the "control mask". Then the array assignment statements are executed in order such that the expression on the right-hand side of each array assignment statement is evaluated and assigned to the array variable on the left-hand side only for those array element positions corresponding to *true* elements in the control mask. For those array element positions corresponding to *false* elements in the control mask, neither an assignment nor an expression evaluation is performed. The elemental assignments are executed according to the rules for (scalar) assignment statements.

```
real, dimension (100)           :: xarray
integer, dimension (100)        :: sparse
logical, dimension (100), save :: mask = .false.
mask(1:100:2) = .true.
where (mask)
  xarray = xarray - 1.0
  sparse = 1
end where
```

Every second element of array **xarray** is reduced by 1.0 beginning with the first array element. And the corresponding elements of array **sparse** are marked (i. e., defined) with the value 1. The other elements of the arrays are not redefined.

An ELSEWHERE statement may be used to separate two assignment blocks within the body of a WHERE construct. After execution of the first assignment block for those array element positions corresponding to *true* values in the mask expression, the second assignment block is executed *exactly* for those array element positions corresponding to *false* values in the mask expression.

```
WHERE (mask_expression)
    array_variable₁ = expression₁
    array_variable₂ = expression₂
    ⋮
ELSEWHERE
    array_variableₖ = expressionₖ
```

array_variable$_{k+1}$ = **expression**$_{k+1}$
$\vdots$

END WHERE

Both assignment blocks may be empty.

During execution of this WHERE construct, the mask expression is evaluated first; its result is the control mask. Then the array assignment statements of the where-block are executed in order such that each array assignment statement is executed only for those array element positions corresponding to *true* array elements in the control mask. And finally, the array assignment statements of the elsewhere-block are executed in order such that each array assignment statement is executed only for those array element positions corresponding to *false* array elements in the control mask.

```
real, dimension (12, 31)    :: pressure, temperature
integer, dimension (12, 31) :: rpm, minutes
:
where (temperature <= 100.0)
  pressure  = 0.99*pressure
  rpm = rpm - 100
elsewhere
  pressure  = 1.01*pressure
  rpm       = rpm + 100
  minutes   = minutes + 1
end where
```

The array assignment statements within the where-block are executed for those array element positions where the temperature is not greater than 100. And the array assignment statements within the elsewhere-block are executed only for those array element positions where the temperature is greater than 100.

8.4.2 Common Rules for Masked Array Assignments

An assignment block within the body of a WHERE construct is allowed to contain only intrinsic array assignment statements.

The statements within a WHERE construct are executed in sequence.

Function reference: If the expression in the right-hand side or the array variable in the left-hand side of an array assignment statement within a WHERE construct includes a reference to a *nonelemental* function, then the actual argument expressions in this function reference and the function itself are evaluated completely without consideration of the mask expression. And then, if the result of the function is an array and the reference is not within the argument list of a nonelemental function, only those array elements are selected for use in evaluating the expression which correspond to *true* values in current control mask.

If the expression in the right-hand side or the array variable in the left-hand side of one of the array assignment statements includes a reference to an elemental intrinsic function as an *operand* (but not within the actual argument list of a reference to a nonelemental function), then this function is processed only for those elements corresponding to *true* elements of the current control mask.

If an array constructor appears in an array assignment statement or in a mask expression within WHERE construct, the array constructor is evaluated without consideration of any masked control and then the array assignment statement is executed and the mask expression is evaluated, respectively.

```
real, dimension (10, 10) :: x, y
  ⋮
where (x > 0.0)
  y = log(x)
  ⋮
elsewhere
  x = -x + 1.0e-3
  y = product( log(x)) - 1
end where
```

The reference to the elemental LOG function in the where-block is evaluated element-wise for all positive array elements of x. But in the elsewhere-block, log(x) is evaluated for the complete array x, because, in this case, log(x) is not an operand on the right-hand side of the assignment statement. PRODUCT is another nonelemental intrinsic function.

Side effects: The assignment to an array variable in the left-hand side of an array assignment statement which also appears as a variable within the mask expression has no effect on the current control mask. A function reference within the mask expression is allowed to (re)define variables appearing also in an array assignment statement within the WHERE construct.

Note that an array assignment statement within the body of a WHERE construct may affect subsequent contained array assignment statements.

9 DECLARATIONS AND SPECIFICATIONS

Specification statements are nonexecutable statements. They are used, for example, to specify properties of data entities and subprograms or to specify the accessibility of certain entities in modules. Every data entity has a type, a rank and, if necessary, other properties. Most of these properties are called *attributes*. If a data entity has a name, that is, if it is a named variable, a named constant, or a function result, its attributes must be specified in the type declaration statement of the data entity.

There are the following *type declaration statements*:

$$
\left.
\begin{array}{l}
\text{INTEGER} \\
\text{REAL} \\
\text{COMPLEX}
\end{array}
\right\}
\begin{array}{l}
\text{for} \\
\text{numeric} \\
\text{data types}
\end{array}
\left.
\begin{array}{l}
\\
\text{for} \\
\text{intrinsic} \\
\text{data types}
\end{array}
\right.
$$

LOGICAL

CHARACTER

TYPE   for derived types

And there are the following *additional specification statements*:

  IMPLICIT   INTRINSIC   PRIVATE   PUBLIC

A specification statement consists of a statement keyword (that is, the name of the statement) and an additional list of those entities for which certain properties are to be specified. Except for the PRIVATE and PUBLIC statements, these lists must not be empty.

A particular attribute may be specified only once for a named entity in a scoping unit. And the name of the entity may appear only in one particular specification statement. Different named entities with the same attribute(s) may be distributed over several specification statements; for example, a specification part may include several INTEGER statements.

```
integer :: ug, og, mine
character (len=1), dimension (72) :: z, line
```

These specifications may also be written as follows:

```
integer :: ug
integer :: og
integer :: mine
character (len=1), dimension (72) :: z
character (len=1), dimension (72) :: line
```

The declaration of different data entities and the specification of the attributes
of a particular data entity may appear in any order, except when already declared entities are used afterwards for the declaration or for the specification of
attributes of other data entities.

```
integer, parameter  :: two = 2
real (kind= two)    :: cc
integer, parameter  :: kind_cc = kind(cc)
real (kind_cc)      :: dd
```

The type declaration for variable cc must appear before the type declaration
for variable dd because the kind type parameter of cc is used within the type
declaration for dd.

9.1   Attributes

Each data entity has properties such as a data type and a rank. In addition, a
data entity may have other properties such as a name, a kind type parameter,
a length, a shape, an initial value, etc.

Except for the type, the kind type parameter, the character length, and the
name, these properties are **attributes**. Some attributes may be specified for
entities that are no data entities.

Attributes for a named data entity must be specified explicitly in its type declaration statement.

Except for the initial value attribute, the name of an attribute is equal to the
keyword which is used to specify the attribute in a type declaration statement.

Attribute	Use
ALLOCATABLE	allocatable array
Initial value	initialization of a variable
DIMENSION	rank and shape of an array
INTENT	intended use of dummy argument
OPTIONAL	optional dummy argument
PARAMETER	named constant expression
POINTER	pointer
PRIVATE	never available outside a module
PUBLIC	may be available outside a module
SAVE	retains certain properties of local variables in subprograms
TARGET	a pointer may point at a target

9.1.1 ALLOCATABLE Attribute

The ALLOCATABLE attribute is used to specify that an array is an *allocatable array*.

`real, allocatable, dimension (:, :) :: aa2, aa22, aa3`

When the ALLOCATABLE attribute is specified, the following rules apply:

The ALLOCATABLE attribute may be specified only for arrays. The array specification must not include array bounds.

The ALLOCATABLE attribute must not be specified for dummy arguments or function results.

An array that has the ALLOCATABLE attribute must not have an additional POINTER or PARAMETER attribute.

9.1.2 Initial Value

Initial values may be specified for nonpointer variables. This kind of initialization is described as "explicit initialization". An initialized variable is already defined with a valid value when the execution of the program begins. It may be redefined during program execution.

`integer, save :: start = 1, border = 0`

When a variable is initialized, the following rules apply:

The assignment of the initial value to a variable is done according to the rules for intrinsic assignment statements.

A variable or a part of it may be explicitly initialized only once in a program.

Within the scoping unit of an interface block for a dummy subprogram or an external subprogram, the specification of an initial value is allowed but the specification is ineffective.

The following entities must not be explicitly initialized: a dummy argument, a function result, an allocatable array, an automatic variable, and a local variable within the specification part of a main program or a function.

If an initial value is specified, the SAVE attribute must also be specified. Thus initialized variables always have the SAVE attribute.

9.1.3 DIMENSION Attribute

The DIMENSION attribute is used to specify that a data entity is an array.

`real, dimension (2:8, 100) :: market, fruit`

9.1.4   INTENT Attribute

The INTENT attribute is used to specify the intended use of dummy arguments.

There are the following INTENT attributes:

INTENT(IN):   The INTENT(IN) attribute specifies that the dummy argument is intended for use as an *input argument*, that is, to receive values from the invoking (sub)program. Such a dummy argument may be used (referenced) but must neither be redefined nor become undefined during the execution of the program.

INTENT(OUT):   The INTENT(OUT) attribute specifies that the dummy argument is intended for use as an *output argument*, that is, to return values to the invoking (sub)program. Such a dummy argument may be used (referenced) only if it has (gotten) a valid value. An associated actual argument must be a variable.

INTENT(INOUT):   The INTENT(INOUT) attribute specifies that the dummy argument is intended for use as an *input/output argument*. Such a dummy argument may receive values from the invoking (sub)program and may return values to the invoking (sub)program. An associated actual argument must be a variable.

```
integer, intent (in) :: in1, in2
real, intent (out)   :: out
```

When an INTENT attribute is explicitly specified, the following rules apply:

An INTENT attribute must be specified for each dummy argument that is a nonpointer variable.

A dummy argument which has an INTENT attribute must not have an additional POINTER attribute.

If a dummy argument of a derived type has an INTENT attribute, its components have (implicitly) the same INTENT attribute.

9.1.5   OPTIONAL Attribute

The OPTIONAL attribute is used to specify for a dummy argument that the subprogram reference need not contain an actual argument corresponding to the dummy argument. A dummy argument with OPTIONAL attribute is an "optional" dummy argument.

```
real, optional :: da1, da2, da3
```

The OPTIONAL attribute may be specified for dummy arguments representing variables or functions. It cannot be specified for a dummy subroutine.

9.1.6 PARAMETER Attribute

The PARAMETER attribute is used to specify a name to be a **named constant**. A named constant is a name for the result of an *initialization expression*. Such a named constant may be used everywhere a literal constant of the same type may be used (exceptions see below).

```
real, parameter :: r = 2.5, pi = 3.1415, d = 22.0/(2*pi)
```

When the PARAMETER attribute is specified, that is, if a named constant is defined, the following rules apply:

The assignment of the initialization expression is done according to the rules for intrinsic assignment statements.

The PARAMETER attribute may be specified only for data entities.

The PARAMETER attribute must not be specified for dummy arguments, function results, allocatable arrays, and pointers.

If the PARAMETER attribute is specified for a data object of a derived type, the components of the object also (implicitly) have the PARAMETER attribute.

A named constant must not have an additional POINTER, TARGET, or SAVE attribute.

A variable must not have the PARAMETER attribute.

The length specification for a named character constant may be "LEN = *".

Another named constant may appear in the initialization expression for the definition of a named constant:

- If the other named constant is already defined lexically earlier in the same type declaration statement statement;
- If the other named constant is already defined lexically earlier in another type declaration statement; or
- If the other named constant is already accessible by USE association or host association.

A named constant must be used neither in a format specification nor as a part of another numeric constant.

9.1.7   POINTER Attribute

The POINTER attribute is used to specify that a data entity is a *pointer*.

```
real, pointer :: coeff, inter, result
```

When the POINTER attribute is specified, the following rules apply:

The pointer attribute may be specified only for variables and for function results.

An entity that has the POINTER attribute must not have an additional ALLOCATABLE, INTENT, PARAMETER, or TARGET attribute.

9.1.8   PRIVATE Attribute

The PRIVATE attribute is used to specify that certain entities in a module cannot be made accessible outside the module. The PRIVATE attribute may be specified for variables, module subprograms, derived data types, named constants, type components, and generic interface blocks. Entities with PRIVATE attribute are described as "private" entities.

```
integer, private :: code, mark, mask
private assignment(=), char_to_num
```

When the PRIVATE attribute is specified, the following rules apply:

The PRIVATE attribute may be specified only in the specification part (i. e. in the scoping unit or in a derived type definition) of a module.

Each entity being declared or defined in the specification part of a module and each module subprogram must be specified either to have the PRIVATE attribute or the PUBLIC attribute.

Entities imported into a module by USE association have the PRIVATE attribute if the module with the USE statement contains any other specification statement. Therefore a PRIVATE statement without a list must appear in the scoping unit of such a module.

9.1.9   PUBLIC Attribute

The PUBLIC attribute is used to specify that certain entities in a module can be made accessible outside the module. The PUBLIC attribute may be specified for variables, module subprograms, derived data types, named constants, type components, and generic interface blocks. Entities with PUBLIC attribute are described as "public" or "visible" entities.

```
real, public :: pr_1, pr_2, pr_10
public operator(.ctof.), celsius_to fahrenheit
```

When the PUBLIC attribute is specified, the following rules apply:

The PUBLIC attribute may be specified only in the specification part (i. e. in the scoping unit or in a derived type definition) of a module.

Each entity being declared or defined in the specification part of a module and each module subprogram must be specified either to have the PUBLIC attribute or the PRIVATE attribute.

Entities imported into a module by USE association have the PUBLIC attribute if the module with the USE statement contains only USE statements. Therefore a PUBLIC statement without a list must appear in the scoping unit of the module.

If a derived data type has the PRIVATE attribute, a data entity of this type must not have the PUBLIC attribute.

9.1.10   SAVE Attribute

The SAVE attribute is used if certain properties of local variables in a subroutine are to be preserved when the END SUBROUTINE statement or a RETURN statement of a subprogram is executed. Variables with SAVE attribute are described as "saved" entities.

If the SAVE attribute is specified in the scoping unit of a subroutine, the following properties of an affected variable are preserved (for the next execution of the subroutine): its definition status, its value, its pointer association status (if it is a pointer), and its allocation status (if it is an allocatable array).

```
real, save :: x, temp, mean
```

When the SAVE attribute is explicitly specified, the following rules apply:

The SAVE attribute may be specified only for variables.

The SAVE attribute must not be specified for dummy arguments, function results, subprogram names, automatic variables, named constants, and local variables in the specification part of a main program and of a function.

An entity that has an explicitly specified SAVE attribute must not have an additional INTENT, OPTIONAL, or PARAMETER attribute.

If an initial value is specified for a variable, an additional SAVE attribute must be specified.

9.1.11 TARGET Attribute

The TARGET attribute is used to specify that a data entity may be associated with a pointer, which then points at this target.

`real, target :: k_array, t, right_part`

When the TARGET attribute is specified, the following rules apply:

An entity that has the TARGET attribute must not have an additional PARAMETER, or POINTER attribute.

The TARGET attribute may be specified only for variables.

If a data object has the TARGET attribute, any subobject of this parent object (implicitly) also has the TARGET attribute. This rule applies to character substrings, for structure components, for array elements, and for array sections without vector subscripts.

Note that an allocatable array may have an TARGET attribute.

9.2 Type Declaration Statements

A type declaration statement is used to explicitly declare one or more data entities. In addition to the data type of a data entity, attributes may be specified such as its value, its accessibility, its rank and its shape, its properties as an actual argument or as a dummy argument, its status as a (local) variable in a subprogram, whether it is a pointer or possibly a target, etc.

type [**, attribute**]... **:: data_entity** [**, data_entity**]...

where **type** is one of the following *type specifications*:

INTEGER	INTEGER (KIND = kind)
REAL	REAL (KIND = kind)
COMPLEX	COMPLEX (KIND = kind)
LOGICAL	LOGICAL (KIND = kind)
CHARACTER (LEN = length)	
TYPE (type_name)	

The **type_name** is the name of a derived type. And **kind** is a is a scalar integer named constant specifying a nonnegative kind type parameter value. The allowed values of the kind type parameter are processor-dependent.

And each **attribute** is one of the following specifications:

ALLOCATABLE	for allocatable arrays,
DIMENSION (...)	for an array specification,
INTENT (IN)	for input arguments,
INTENT (OUT)	for output arguments,
INTENT (INOUT)	for input/output arguments,
OPTIONAL	for *optional* dummy arguments,
PARAMETER	for named constants,
POINTER	for pointers,
PRIVATE	for *private* entities in modules,
PUBLIC	for *public* entities in modules,
SAVE	for *saved* local entities in subprograms, and
TARGET	for pointer targets.

And each **data_entity** is one of the following:

– If it is a variable (also a function result), it may be the name of the variable:

name

– If it is a nonpointer variable (but no function result) or a named constant, it may be the name of the object followed by an equals with a subsequent *initialization expression*:

name = initialization_expression

```
integer, parameter :: four = 4, eight = 8
integer :: number, years
integer (kind= four) :: max, min
real, parameter :: zero = 0.0, one = 1.0
character (len=8), dimension (18:48), save :: sweden, king
type (date), dimension (25, 13) :: student, parents
```

Type declaration

The type of a named data entity must be declared *explicitly* by a type declaration statement.

Type parameters: If no kind type parameter is specified in a type declaration statement for an intrinsic type, this is a declaration for an entity of type *default* integer, *default* real, *default* complex, *default* logical, or character.

Initialization

If an equal with an initialization expression is specified for a data entity in a type declaration statement, either a named constant is defined or a nonpointer variable is explicitly initialized. In the last case, the specified variable on the left-hand side of the equals becomes explicitly initialized with the result of the initialization expression; that is, an initial value is assigned to the variable when the execution of the program begins. The assignment of the initial value to the variable is done according to the rules for intrinsic assignment statements. A variable or a part of it may be explicitly initialized only once in a program.

If an equals with an initialization expression appears, the SAVE attribute must also be specified if the data entity is a variable, and the PARAMETER attribute must be specified if the data entity is a named constant.

An initialization expression must *not* be specified for the following entities: dummy arguments, function results, allocatable arrays, pointers, names of external subprograms, and names of automatic variables.

9.2.1   INTEGER Statement

An INTEGER statement is used to declare named constants, variables and/or function results of type *integer*.

INTEGER [**(KIND = kind)**] [**, attribute**]... **:: data_entity** [**, data_entity**]...

where **kind**, **attribute**, and **data_entity** are defined as on page 9-8/9-9.

```
integer :: a, x2, nn, fe, k
integer, parameter :: long = selected_int_kind(8)
integer (kind= long), dimension (0:2, 11:14, 5), save :: f1, f2
```

Regardless of whether or not a kind type parameter is explicitly specified, the F processor selects (if possible) a suitable processor-dependent method of internal representation of the values of the specified integer data entities.

If a kind type parameter is not explicitly specified, it is a declaration of entities of type *default integer* and the F processor uses the kind type parameter value KIND(0).

9.2.2   REAL Statement

A REAL statement is used to declare named constants, variables and/or function results of type *real*.

REAL [**(KIND = kind)**] [**, attribute**]... **:: data_entity** [**, data_entity**]...

where **kind**, **attribute**, and **data_entity** are defined as on page 9-8/9-9.

```
real :: measure, value, tax, direction, height
real, dimension (2), save :: e = (/1.2,0.3/), f = (/(7.5, i=1,2)/)
```

```
real, parameter :: pi = 3.1415, e = 2.7182
  integer, parameter :: ten = 10
real (kind= ten) :: a, b, c, d
```

Regardless of whether or not a kind type parameter is explicitly specified, the F processor selects (if possible) a suitable processor-dependent approximation method for the internal representation of the values of the specified real data entities.

If a kind type parameter is not explicitly specified, it is a declaration of entities of type *default real* and the F processor uses the kind type parameter value KIND(0.0).

9.2.3   COMPLEX Statement

A COMPLEX statement is used to declare named constants, variables and/or function results of type *complex*.

COMPLEX [**(KIND = kind)**] [**, attribute**]... **:: data_entity** [**, data_entity**]...

where **kind**, **attribute**, and **data_entity** are defined as on page 9-8/9-9.

```
complex :: point, vector, two, plane, cc
complex, private, dimension (100) :: c1, c_denominator, circle
  integer, parameter :: low_prec = kind(0.0)
complex (kind= low_prec) :: c1, c2
```

Regardless of whether or not a kind type parameter is explicitly specified, the F processor selects (if possible) a suitable processor-dependent approximation method for the internal representation of the values of the real and imaginary parts of the specified complex data entities.

If a kind type parameter is not explicitly specified, it is a declaration of entities of type *default complex*, and the F processor uses the same kind type parameter value as for entities of type default real, namely KIND(0.0).

High precision complex entities: If **kind** is the kind type parameter value of a real type with increased precision, the approximation method for the internal representation of the real and imaginary parts of the specified entities is the same as that of corresponding real entities with the same increased precision.

9.2.4   LOGICAL Statement

A LOGICAL statement is used to declare named constants, variables and/or function results of type *logical*.

LOGICAL [**(KIND = kind)**] [**, attribute**]... **:: data_entity** [**, data_entity**]...

where **kind, attribute**, and **data_entity** are defined as on page 9-8/9-9.

```
logical :: my, lambda, omega, switch, wf
logical, dimension (-n:n, 1987:1988) :: s1, jn, hit
  integer, parameter :: one = 8, bit = 1
logical (kind= one), allocatable, dimension (:) ::  pattern
logical (kind= bit), dimension (10) :: pattern1, pattern2
```

Regardless of whether or not a kind type parameter is explicitly specified, the F processor selects (if possible) a suitable processor-dependent method for the internal representation of the values of the specified logical data entities.

If a kind type parameter is not explicitly specified, it is a declaration of entities of type *default logical* and the F processor uses the kind type parameter value KIND(.FALSE.).

9.2.5 CHARACTER Statement

A CHARACTER statement is used to declare named constants, variables and/or function results of type *character*.

CHARACTER (LEN = length) [**, attribute**]... **:: data_entity** [**, data_entity**]...

where **attribute** and **data_entity** are defined as on page 9-8/9-9. And **length** is either a specification expression or an asterisk *.

```
character (len=3), parameter :: one = "one", two = "two"
character (len=8), dimension (100) :: abcfld, atom, text
character (len= n+2) t1, t2, t3, t4
```

9.2.5.1 Length Specification

The character **length** indicates the number of characters in the declared character entities.

The character length may be zero. If the result of the specification expression **length** is negative, the character length zero is used. The length specification for an array object is interpreted as the length of its single array elements.

The length specification for a character entity may contain a specification expression; in a simple case, it is an integer literal constant without a sign. If the length specification appears in a main program, the specification expression must be a scalar integer *constant expression*.

If the specification expression in a length specification is *not* a constant expression, the length is determined immediately before the execution of the subprogram. And a later redefinition of any operand in the specification expression has no effect on the current length, as far as this invocation and execution of the subprogram is concerned. The length will be reevaluated only when the subprogram is invoked the next time. Note that if this nonconstant length is specified for a variable which is no dummy argument, it is an *automatic* character variable.

Length specifications LEN = *

If an asterisk * appears in a length specification instead of a specification expression, this specifies either the length for a dummy argument or the length for a named constant.

Dummy argument: An * must be specified as character length of a dummy argument of type character; this form is called "assumed length".

Named constant: If the length * is specified for a named constant, the named constant automatically has the length of the result of the initialization expression defining the named constant.

```
character (len=*), parameter :: j = "January", m = "March"
```

The named constants j and m have the lengths 7 and 5, respectively.

9.2.6   TYPE Declaration Statement

A TYPE declaration statement is used within a module to declare named constants, variables and/or function results of a *derived type*.

TYPE (type_name) [**, attribute**]... **:: data_entity** [**, data_entity**]...

where **type_name** is the name of an available derived type. And **attribute** and **data_entity** are defined as on page 9-8/9-9.

A declaration of derived type entities is only possible either inside a module if the type definition for this derived type appears lexically earlier in the module or outside the module containing the type definition if the derived type definition is there accessible by USE association.

```
type, public :: page1
  character (len=60), dimension (125) :: p
end type page1

type, public :: page2
  character (len=80), dimension (25) :: p
end type page2
```

In a scoping unit where these derived type definitions are available (i. e., specified or accessible), the following declarations may be written:

```
type (page1) :: printer1, printer2
type (page2) :: screen
```

9.3   Additional Specification Statements

9.3.1   PRIVATE Statement

The PRIVATE statement may be used to specify the PRIVATE attribute for certain entities in a module.

The PRIVATE statement is a nonexecutable statement which may appear only in the specification part of a module (within the scoping unit of the module or within derived type definitions).

PRIVATE

PRIVATE :: module_entity [, module_entity]...

Each **module_entity** is the specific name of a module subprogram, the generic name of a module subprogram, OPERATOR(**operator**) designating a (generic) operator interface block for the specified **operator**, or ASSIGNMENT (=) designating a (generic) assignment interface block.

The PRIVATE statement *without* a list of module entities specifies

1. in the scoping unit of a module that all those module entities that are imported from other modules are private (i. e. cannot be made accessible outside the module with the PRIVATE statement). This statement must appear in a private module which contains at least one USE statement and one ore more other specification statements (in addition to the PRIVATE statement).

2. in the scoping unit of a derived type definition within the specification part of a module that no component of the derived type can be made accessible outside the module.

A module may contain more than one PRIVATE statement. They may appear in the scoping unit of the module and in any derived type definition. But only one such PRIVATE statement without a list may appear in the scoping unit of a module and within each of the derived type definitions.

```
module mol
use data4mol
private
private :: operator(.dist.), root
public  :: a, b, c
```

The module entities imported from `data4mol` have the PRIVATE attribute in module `mol`. All other module entities in module `mol` must be explicitly appear in a PUBLIC or PRIVATE statement. The module entities `a`, `b`, and `c` are visible. The operator `.dist.` and the module entity `root` cannot be made accessible outside module `mol`.

9.3.2 PUBLIC Statement

The PUBLIC statement may be used to specify the PUBLIC attribute for certain entities in a module.

The PUBLIC statement is a nonexecutable statement which may appear only in the specification part of a module (within the scoping unit of the module).

PUBLIC

PUBLIC :: module_entity [, module_entity]...

where each **module_entity** is defined as for the PRIVATE statement.

The PUBLIC statement *without* a list of module entities specifies in the scoping unit of a module that module entities imported from other modules are visible (i. e. can be made accessible outside the module). This statement must appear in a module which contains only one or more additional USE statements and no other specification statement.

A module subprogram with a dummy argument of a private type or a module function with a private result must be specified to have the PRIVATE attribute and must not have an additional generic name with PUBLIC attribute.

A private module may contain more than one PUBLIC statement. But if a module contains a PUBLIC statement *without* a list, the other statements in the module must be USE statements only.

```
module m
use data4m
private
public  :: assignment(=), circle
private :: x, y, z
```

The module entities imported from **data4m** have the PRIVATE attribute in module **m**. All other module entities in module **m** must be explicitly appear in a PUBLIC or PRIVATE statement. The module entities **x**, **y**, and **z** are private. The defined assignment and the module entity **circle** can be made accessible outside module **m**.

9.3.3 IMPLICIT Statement

The IMPLICIT statement is used in Fortran to specify implicit typing for those named data entities not being explicitly declared in a type declaration statement. Only to improve the portability of F programs, the following form of the IMPLICIT statement is allowed in the main program and in the specification part of a private module (i. e., a module which contains not only USE statements):

IMPLICIT NONE

In the F language, this statement is unneccessary and is ignored because even without an IMPLICIT NONE statement each named data entity must be declared explicitly.

9.3.4 INTRINSIC Statement

The INTRINSIC statement may be used to specify a name as the name of an intrinsic subprogram.

INTRINSIC name [, name]...

Each **name** is the name of an intrinsic subprogram.

```
intrinsic sin, sqrt, exp
```

The names **sin**, **sqrt**, and **exp** are treated in the scoping unit with the INTRINSIC statement as the names of intrinsic functions.

10 EXECUTION CONTROL

Normally, the statements in a main program or subprogram are executed line by line in the order in which they appear in the source program. Execution control statements are used to alter or to terminate this sequential execution sequence. The execution sequence is called the *control flow*. Any change of the sequential execution sequence is a **branching** or **transfer of control**. There are *simple* execution control statements, and there are execution control *constructs* consisting of several statements.

The following execution control statements are **simple** executable statements: CALL, END PROGRAM, END SUBROUTINE, END FUNCTION, RETURN, and STOP, and finally EXIT and CYCLE, which may appear only in DO constructs.

The following execution control statements are used to form *constructs*:

- DO and END DO form a DO construct (that is, a DO loop);

- IF THEN, ELSE IF, ELSE, and END IF form an IF construct; and

- SELECT CASE, CASE, and END SELECT form a CASE construct.

DO constructs, IF constructs, and CASE constructs are compound statements consisting of one or more statement sequences. Such a sequence of statements is called a **block**. A construct has an internal control flow, which automatically branches within the construct.

If the control flow depends on certain values, these values must be scalar.

10.1 IF Construct

An IF construct may contain one or more blocks. It controls the execution of at most one of its constituent blocks. The conditional execution of such a block may depend on one or more conditions.

The first statement of an IF construct is an IF THEN statement. And the last statement of an IF construct is an END IF statement. Additional ELSE statements and ELSE IF statements may be used to control the control flow within the IF construct.

```
if (colour == "red") then        ! IFTHEN statement
  :                              ! if-block
elseif (colour == "green") then  ! corresponding ELSEIF stmt
  :                              ! elseif-block
else                             ! corresponding ELSE statement
  :                              ! else-block
endif                            ! corresponding ENDIF stmt
```

An IF THEN statement is the first statement of an IF construct.

IF (logical_expression) THEN

The **logical_expression** must be scalar.

An ELSE statement is the first statement of an alternative block, which may be conditionally executed after the execution of the corresponding IF THEN statement or after the execution of a corresponding ELSE IF statement. An IF construct may include not more than one corresponding ELSE statement:

ELSE

An ELSE IF statement combines the effect of an IF THEN statement and an ELSE statement. An ELSE IF statement is the first statement of an alternative block, which may be conditionally executed after the execution of the corresponding IF THEN statement or after the execution of a preceding ELSE IF statement.

ELSE IF (logical_expression) THEN

The **logical_expression** must be scalar.

An END IF statement is the last statement of an IF construct. A corresponding END IF statement must appear for each IF THEN statement. The execution of an END IF statement has no effect. An END IF statement has the following form:

END IF

10.1.1   Simple IF Constructs

The basic form of an IF construct containing only one block is:

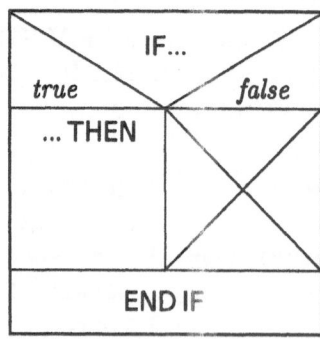

This IF construct controls the conditional execution of one block.

IF (logical_expression) THEN

⋮   ⟵ if-block

END IF

If the **logical_expression** is *true*, program execution continues with the first
executable statement in the if-block; otherwise program execution continues
with the corresponding END IF statement.

```
if (colour == "red") then
  time = 30
  call red(time)
endif
```

If the value of the scalar character variable `colour` is `"red"`, the two statements
in the if-block are executed; otherwise they are ignored.

The basic form of an IF construct containing two alternative blocks is:

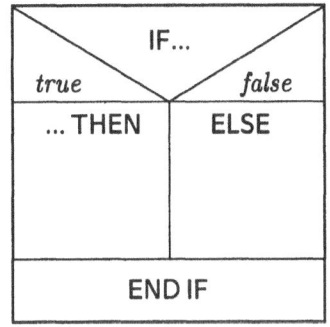

10

Exactly one of the two blocks of this IF construct is conditionally executed.

IF (logical_expression) THEN

⋮   ⟵ if-block

ELSE

⋮   ⟵ else-block

END IF

If the **logical_expression** is *true*, program execution continues with the first
executable statement in the if-block; otherwise program execution continues
with the first executable statement in the else-block. When the execution of the
block is completed, program execution continues with the corresponding END IF
statement.

```
if (colour == "red") then
  time = 30
  call red(time)                    ! if-block
else
  time = 20
  call green(time)                  ! else-block
endif
```

If the value of the scalar character variable `colour` is `"red"`, the two statements in the if-block are executed, the two statements in the else-block are ignored, and program execution continues with the corresponding END IF statement. But if the character variable `colour` does *not* have value `"red"`, the two statements in the if-block are ignored, the ELSE statement, the two statements in the else-block, and finally the corresponding END IF statement are executed.

The basic form of an IF construct containing two alternative blocks, where the first block depends on one condition and the second block depends on two conditions, is:

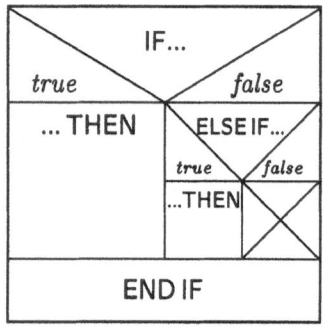

At most one of the two blocks of this IF construct is conditionally executed.

An IF construct may include more than one ELSE IF statement if the execution of blocks depends on more than one condition. Note that such a form of an IF construct is not a nested IF construct:

IF (logical_expression) THEN
⋮   ⟵ if-block
ELSE IF (logical_expression) THEN
⋮   ⟵ first elseif-block
ELSE IF (logical_expression) THEN
⋮   ⟵ second elseif-block
END IF

If the logical expression in the IF THEN statement is *true*, program execution continues with the first executable statement in the if-block; otherwise program execution continues with the first ELSE IF statement. If the logical expression in the first ELSE IF statement is *true*, program execution continues with the first executable statement in this elseif-block; otherwise the second ELSE IF statement is executed. If the logical expression in the second ELSE IF statement is *true*, program execution continues with the first executable statement in this elseif-block; otherwise program execution continues with the corresponding END IF statement. When the execution of the selected block is completed, program execution continues with the corresponding END IF statement.

An IF construct containing more than two corresponding ELSE IF statements is analogously executed.

An IF construct with ELSE IF statements may also include an ELSE statement. This form of an IF construct is different from that of the last one as follows: when the logical expressions in the IF THEN statement and in all ELSE IF statement(s) are *false*, program execution continues with the first executable statement in the else-block.

```
if (colour == "red") then
  time = 30
  call red(time)
elseif (colour == "green") then
  time = 20
  call green(time)
elseif (colour == "yellow") then
  time = 5
  call yellow(time)
else
  call error()
endif
```

The ELSE statement causes one block of the IF construct to be executed in any case.

10.1.2 Nested IF Constructs

IF constructs may be nested; that is, an IF construct may appear within any block of another IF construct.

If the inner IF construct is in the else-block or in one of the elseif-blocks of the outer IF constructs, this is a piece of program which may also be written without nesting IF constructs by the use of an additional ELSE IF statement.

```
if (c == "red") then          ! if (c == "red") then
   time = 30                   !    time = 30
   call red(time)             !    call red(time)
else                          ! elseif (f == "green") then
   if (f == "green") then     !
      time = 20               !    time = 20
      call green(time)        !    call green(time)
   else                       ! elseif (c == "yell") then
      if (c == "yell") then   !
         time = 5             !    time = 5
         call yellow(time)    !    call yellow(time)
      endif                   !
   endif                      !
endif                         ! endif
```

The nested construct on the left-hand side is equivalent to the nonnested construct on the right-hand side.

If the inner IF construct is in the if-block of the outer IF construct, the execution of the if-block of the inner IF construct depends on both the logical expression in the outer IF THEN statement and the logical expression in the inner IF THEN statement.

An inner IF construct must be entirely contained within the if-block, within the else-block, or within one of the elseif-blocks of the outer IF construct.

10.2 CASE Construct

A CASE construct selects for execution at most one of its constituent blocks. The first statement of a CASE construct is a SELECT CASE statement. The last statement of a CASE construct is an END SELECT statement. And the CASE and CASE DEFAULT statements are used to control the control flow within the CASE construct. The CASE DEFAULT statement is a special case of a CASE statement.

```
select case (colour)   ! SELECTCASE statement
case ("red")           ! corresp. CASE statement with selector
   :                   ! first case-block, case "red"
case ("green")         ! corresp. CASE statement with selector
   :                   ! second case-block, case "green"
case default           ! corresp. CASE stmt. with selector DEFAULT
   :                   ! third case-block, case 'otherwise'
end select             ! corresponding ENDSELECT statement
```

Normally, a CASE construct consists of several alternative blocks. Such a block is called a **case-block**. The execution of the SELECT CASE statement causes the *case expression* to be evaluated; its result is called the **case index**. Then, this case index is used to determine whether or not any case-block in the CASE construct is to be executed or which case-block is to be executed. If a CASE construct includes a CASE DEFAULT statement, *exactly* one case-block is selected for execution.

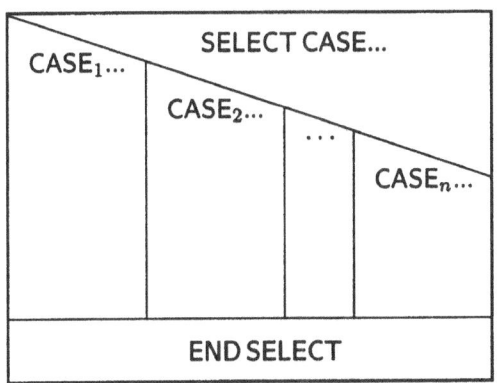

Each "case" in a CASE construct consists of a sequence of executable statements, a case-block. The default-block is a special case-block.

A CASE construct may include one, none, or more than one corresponding CASE statements. A case-block or the default-block of a CASE construct may be empty.

SELECT CASE (case_expression)
CASE selector
  ⋮   ⟵ first case-block
CASE selector
  ⋮   ⟵ second case-block, etc.
CASE DEFAULT
  ⋮   ⟵ default-block
END SELECT

The SELECT CASE statement containing the case expression is the first statement of a CASE construct.

SELECT CASE (case_expression)

The **case_expression** is a scalar integer or character expression.

```
select case (n*3 - j)            ! integer
select case ("light " // colour) ! character
```

A CASE statement is the first statement of a case-block. It has the following form:

CASE selector

The **selector** designates the **case values** for the selection of the *case-block* following the CASE statement.

The form of the **selector** is either **DEFAULT** or **(cv [, cv]...)**, where each **cv** is a single *case value* or a *case value sequence*. The form of a **case value sequence** is as follows:

case_value :
: case_value
case_value : case_value

Each case value included in a selector must be a scalar *initialization expression* of type integer or character.

The colon notation is interpreted as a sequence of single case values:

Selector	Corresponding *case values*
case_value$_1$: case_value$_2$	**case_value$_1$** $\leq$ *case value* $\leq$ **case_value$_2$**
case_value$_1$:	**case_value$_1$** $\leq$ *case value*
: case_value$_2$	*case value* $\leq$ **case_value$_2$**

DEFAULT is a special selector. If any, it must be the final selector of its construct. The selector DEFAULT matches all values of the case expression for which there is no matching case value in any other CASE statement in the CASE construct.

The type and, if applicable, the kind type parameter of each case value in a particular CASE construct must be those of the case expression. If the case expression is of type character, the lengths of the case values may be different from that of the case expression.

```
case (11)
case (n/2 +1)
case (1,8,10:12,19)              ! case (1, 8, 10, 11, 12, 19)
case (:1933, 1939:1942, 1945:)
case ("one", "two", "three")
case default
```

The case values of the selector in a single CASE statement must *not* overlap. And the case values of the selectors within a particular CASE construct must *not* overlap; that is, for a particular result of the case expression, there must be at most one matching case value in all CASE statements within a CASE construct.

The END SELECT statement is the last statement of a CASE construct. The statement has the following form:

END SELECT

Execution

The execution of a CASE construct causes the scalar case expression to be evaluated. When the case index matches a case value belonging to a selector in any CASE statement within this CASE construct, the case-block following this CASE statement is selected for execution. The execution of this selected case-block completes the execution of the CASE construct. If the case index does not match a case value, program execution continues with the next executable statement following the END SELECT statement.

10.2.1   Simple CASE Constructs

A simple CASE construct includes exactly one CASE statement and one case-block. The conditional execution of the block depends on one condition.

```
select case (n)              ! if (n == 1) then
case (1)                     !
  o = 2*pi*r*h + pi*r**2     !   o = 2*pi*r*h + pi*r**2
  v = pi/2*r**2*h            !   v = pi/2*r**2*h
end select                   ! endif
```

In this case, the CASE construct and the IF construct are equivalent.

The following example implements actions for every meaningful and meaningless value of the case expression. The second case-block is empty. And the third case-block is the default-block, which will be executed when **time** is out of a meaningful range:

```
select case (time)
case (8:18)
  call bell(time)
case (1:7, 19:24)
case default
  call error()
end select
```

10.3 DO Construct

A DO construct is a sequence of statements which may be executed more than once. The first statement of a DO construct is a DO statement. The subsequent block is the "body" of the DO construct. The last statement of a DO construct is an END DO statement. The body may contain CYCLE statements and EXIT statements, which are used to control the control flow within the range of the DO construct.

There are two sorts of DO constructs: *count loops* and *endless loops*. Their general form is:

[**do-name :**] **DO** [...]
  ⋮   ⟵ body of the DO construct
END DO [**do-name**]

If the DO statement includes a **do-name**, the corresponding END DO *must* include the same name, and additional CYCLE and EXIT statements of this construct, if any, *may* include this name. The **do-name** must not be the same as the name of any available entity in the scoping unit. If the DO statement does not include a **do-name**, neither the END DO statement nor the CYCLE and EXIT statements must include a **do-name**.

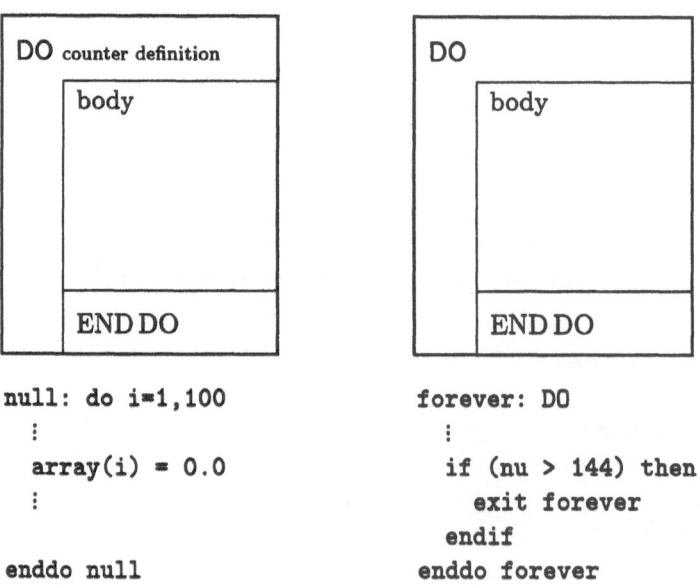

```
null: do i=1,100          forever: DO
   ⋮                         ⋮
   array(i) = 0.0           if (nu > 144) then
   ⋮                           exit forever
                            endif
enddo null               enddo forever
```

The body of a DO construct is also called the "range" of the DO construct.

10.3.1   DO Statement

The execution of a DO statement causes the internal control flow of the DO construct to be initiated.

[do-name :] DO

[do-name :] DO do_variable = first, last [, stride]

The **do-name** identifies the DO construct. **do_variable** is the name of a scalar variable of type integer which must not be a dummy argument, a pointer, a function result, or accessed by USE or host association. And **first**, **last**, and **stride** are scalar integer expressions.

Loop parameters: The scalar results of **first**, **last**, and **stride** are converted, if necessary, to the kind type parameter of the DO variable. These values are called the **initial parameter**, **terminal parameter**, and **increment parameter**, respectively, of the DO construct. The value of the optional **stride** must be nonzero. If the **stride** is absent, the value one is assumed as the default increment parameter.

```
do
do l=i,i+n
inner: do m=1,75,2
```

10

10.3.2   END DO Statement

The termination statement of a DO construct has the following form:

END DO [do-name]

The **do-name** identifies the DO construct. If this name is present, the same name must be included in the DO statement.

10.3.3   Forms of DO Constructs

If the first statement of a DO construct is a DO statement with a DO variable, it is a "count loop". A count loop is controlled by an *iteration count* (see below), which automatically terminates the execution of the DO construct when the maximum limit for the loop iterations is reached.

If the first statement of a DO construct is a DO statement without a DO variable, it is an "endless loop". The execution of such a DO construct can be terminated only by a terminating statement within its range.

```
do i=1,100              ! count loop, no do-name
  odd(i) = 2*i + 1
enddo                   ! no do-name


maximum: do             ! endless loop, do-name
  if(x > y) then
    xmax = x
    exit maximum        ! exit this loop
  endif
  read *, x
enddo maximum           ! do-name
```

10.3.4   Execution of a DO Construct

When a DO construct is executed, the following steps are performed: loop control is initiated, then the range is executed (generally several times), and finally, loop execution is terminated.

Loop control initiation

Count loop only: First, the loop parameters are determined (10.3.1). Suppose, $m1$ is the value of the initial parameter, $m2$ is the value of the terminal parameter, and $m3$ is the value of the increment parameter. Then the DO variable is defined with $m1$. Finally, the *iteration count* is evaluated:

$$\text{Iteration count} = \text{MAX}\,((m2 - m1 + m3)/m3,\, 0)$$

Execution cycle

1. Count loop only: The iteration count is tested. If the iteration count is greater than zero, program execution continues with the first (executable) statement in the range. If the iteration count is zero, the count loop is "satisfied" and the execution of the DO construct terminates (see below).

2. Beginning with the first executable statement in the range, (executable) statements are executed until the END DO statement is executed, until a CYCLE statement is executed, or until the loop is terminated during this execution cycle.

3. Count loop only: The iteration count is decremented by one, and the DO variable is incremented or decremented, depending on the sign of the increment parameter $m3$, by the absolute value of the increment parameter.

4. The next iteration of the loop is executed (beginning with step 1).

These steps are repeated until the execution of the loop terminates (see below).

Loop termination

The execution of a DO construct terminates:

- When the iteration count is zero.
- When a RETURN statement is executed in the range.
- When an EXIT statement is executed in the range.
- When an EXIT or CYCLE statement of a surrounding DO construct is executed in the range.
- When a STOP statement is executed or when the execution of the program is terminated for any other reason.

When a DO construct (but not the program) terminates, control is transferred to the next (executable) statement following the END DO of the terminated DO construct.

10

10.3.4.1   Additional Details about Count Loops

Iteration count: The iteration count may already be zero before the first iteration, for example, if both $m1 > m2$ and $m3 > 0$ are *true*, or if both $m1 < m2$ and $m3 < 0$ are *true*. In such a case, the range is not executed and the DO construct is skipped.

The maximum number of iterations is determined when the DO construct is initiated. This maximum number of iterations cannot be modified within the range. It is equal to the actual number of iterations if the DO construct is not terminated until it is satisfied.

DO variable: When a DO construct terminates before the first iteration is executed, the DO variable retains the value of the initial parameter. When a DO construct terminates before the iteration count is zero, the DO variable retains its last defined value. When the DO construct terminates because the iteration count is zero, the DO variable also retains its last defined value, that is, its value during the last iteration plus the value of the increment parameter $m3$.

Except for the automatic incrementation or decrementation, the DO variable must neither be redefined nor become undefined during the execution of the DO construct.

```
never = 100
do i=never,85
   :
enddo
```

During the execution of the DO statement, the initial parameter is greater than the termination parameter; that is, the iteration count already is zero before the first iteration. Therefore, the range is not executed.

```
back: do k=10,1,-1
  a(k) = 11 - k
enddo back
```

This DO construct is interpreted as:

```
a(10) = 1
a(9)  = 2
a(8)  = 3
:
a(3)  = 8
a(2)  = 9
a(1)  = 10
k     = 0
```

If the increment parameter is negative, the DO variable is not incremented but decremented.

```
do m=1,11,3
  if (f(m) < 0.0) then
    f(m) = - f(m)
  endif
  print *, m, f(m)
enddo
call sub()
```

The range of this DO construct is executed four times because the DO variable becomes defined with 1, 4, 7, and 10. After termination of the DO construct, the DO variable retains the value 13. And program execution continues with the CALL statement.

10.3.4.2 Additional Details about Endless Loops

An endless loop has no upper limit for the number of iterations. A termination condition may appear only in the range.

```
ever: do
  if (age < 100) then   ! termination condition
    exit ever
  endif
  :
enddo ever
```

10.3.4.3   CYCLE Statement and EXIT Statement

The CYCLE statement may be used to terminate the current iteration of the corresponding DO construct (but not the complete loop) immediately.

CYCLE [do-name]

The execution of a CYCLE statement has the effect such as a branch to the END DO statement of a DO construct. Precisely: after the execution of a CYCLE statement, no other statement in the corresponding DO construct is executed during this iteration and the execution of the DO construct continues with the next iteration.

The EXIT statement may be used to terminate the execution of the corresponding DO construct immediately.

EXIT [do-name]

Mostly (as in the following example), the execution of an EXIT statement has the effect such as a branch to the next executable statement subsequent to the corresponding END DO statement of the DO construct. In any case, after the execution of an EXIT statement, no other statement in the corresponding DO construct is executed.

```
year = 0
read *, salary
do
  year = year + 1
  salary = 1.04 * salary
  if (year > 15) then
    exit
  endif
  if (salary > 6500.0) then
    cycle
  endif
  salary = salary + base
enddo
```

10.4   Nested Constructs

IF constructs, CASE constructs, and DO constructs may be nested in each other,
but they must not overlap; that is, one block of such a construct may contain
another complete construct even of the same sort.

```
character, dimension (80):: line
integer :: i, level
level = 0
do i=1,80
  select case (line(i:i))
  case ("(")
    level = level + 1
  case (")")
    level = level - 1
    if (level < 0) then
      print *, - level, " right parentheses excess"
      exit
    endif
  case default     ! only parentheses are checked
  end select
enddo
if (level > 0) then
  print *, level, " right parentheses lack"
endif
```

10.5   STOP Statement

The execution of a STOP statement causes the termination of execution of the
program.

STOP

The execution of a STOP statement causes the program to terminate immediately without any possibility to resume execution.

A main program or subprogram may contain several STOP statements. A control print immediately before a STOP statement may be used to determine the
actual position where program execution has stopped.

```
print *, "Normal program termination"
stop
```

> Note that CALL, END PROGRAM, END SUBROUTINE, END FUNCTION,
> and RETURN statements also cause a transfer of control. See chapter 13.

11 INPUT/OUTPUT

Input statements are used to transfer data from an external or internal file to internal storage; this is called *reading*. **Output statements** are used to transfer data from internal storage to an external or internal file; this is called *writing*. In addition to the data transfer statements, there are input/output statements that inquire about, specify, or modify the properties of a file or of a *unit*. And finally, there are input/output statements which manipulate the external medium such as for the positioning of external files.

Each of the following terms characterizes a special kind of input/output:

Sequential io	= input/output with sequential access
Direct io	= input/output with random access
Formatted io	= input/output with format specifications
Unformatted io	= input/output without format specifications
List-directed io	= io-statements with * as FMT= specifier
Nonadvancing io	= input/output without automatic positioning
Internal io	= input/output from storage to storage

11.1 Records

11

The basic constituent of the F file system is the *record*. A **record** is a sequence of values or a sequence of characters. For example, normally, a line on the screen is a record. A record need not be a physical record. There are three kinds of records: *formatted* records, *unformatted* records, and *endfile* records.

A **formatted record** consists of a sequence of ASCII characters. An F processor may prohibit certain control characters from appearing in a formatted record.

The length of a formatted record is measured in characters. The length is processor-dependent and depends primarily on the number of characters written into the record. The length may be zero.

Formatted records may be read or written only by formatted input/output statements. They may also be produced by means other than F, such as the keyboard of a terminal.

An **unformatted record** consists of a sequence of values in a processor-dependent internal representation. An unformatted record may contain data of any intrinsic or derived type. The length of an unformatted record is measured in processor-dependent units; its length is processor-dependent and depends on the output list when the record is written. The length may be zero.

Unformatted records may be read or written only by unformatted input/output statements.

An **endfile record** may appear only as the last record of a file. An endfile record has no length (from the point of view of the programmer). An endfile record may be written to a file opened for sequential input/output, and it may be recognized during reading. The (internal) representation of an endfile record is processor-dependent.

An endfile record may be (explicitly) written by the execution of an ENDFILE statement, but it may also be written implicitly. If the file is opened for sequential access and if the most recent input/output statement for that file is a data transfer output statement and no positioning statement, an endfile record is written implicitly (that is, automatically) when a REWIND or BACKSPACE statement is executed for the connected unit, when the file is explicitly closed by the execution of a CLOSE statement, when the file is implicitly closed at program termination.

11.2 Files

A **file** is a sequence of records. There are *external files* and *internal files*.

External files are files that exist in a medium external to the program such as a magnetic disk or a tape storage. But an external file also may be a device such as the printer, the screen, or the keyboard of a terminal. Files are normally organized by the operating system environment in which the *program* is embedded.

Internal files are character variables (that is, internal storage of the program) playing the role of a file for internal input/output.

11.3 File Attributes of External Files

An external file may have the following attributes which are partly processor-dependent: the set of allowed names (processor-dependent), the existence, the set of allowed access methods (processor-dependent), the set of allowed forms (processor-dependent), the set of allowed record lengths (processor-dependent), the set of allowed actions, and the position. Certain file attributes are only established if the file is connected to a unit, that is, if it is opened.

11.3.1 File Names

An external file may have a name. This name can appear only in OPEN and INQUIRE statements.

If a unit is not (explicitly) connected to a file by the execution of an OPEN statement *with* FILE= specifier, the F processor possibly generates a processor-dependent file name.

11.3.2 Access Methods

Records may be read and written in sequential order or in random order. There are two access methods: *sequential access* and *direct access*. Input/output data transfer with sequential access is described as *sequential input/output*, and input/output data transfer with direct access is described as *direct input/output*.

External files may be processed with sequential or direct access. Internal files may be processed only with sequential access.

The access method for an external file which is not preconnected is determined when the file is connected to a unit. For the time of this connection, the access method must not be changed.

11.3.2.1 Sequential Access

If a file is processed only with sequential access, the order in which the records are written determines the order in which they may be read; that is, the nth record can be read only after the first record, the second record, and so on, and the $(n-1)$th record have been read.

If the F processor also allows direct access to the file, the order of the records for the sequential access is given by the record numbers which are associated with the records for the direct access.

External sequential files may be connected to a unit by the execution of an OPEN statement. Such explicit connection is not necessary if the operating system performs the connection, such as in the case of *preconnected units*.

11.3.2.2 Direct Access

If an external file is opened for direct access, the order of the records depends on the order of the *record numbers*. In contrary to sequential input/output, the records need not be read or written in any order. For example, the record with record number 3 may be written though the records with the record numbers 1 and 2 have not (yet) been written.

Input/output actions such as the rewinding of a file at its initial point or as the positioning of a file at its terminal point are meaningless for a file which is opened for direct access. An endfile record, if any, is not considered as a part of the file as long as the file is opened for direct access.

There is formatted and unformatted direct input/output. List-directed input/output with direct access and nonadvancing input/output with direct access are not allowed. Internal files cannot be opened for direct access.

If an external file is to be opened for direct access, an OPEN statement must be executed specifying both ACCESS="direct" and the RECL= specifier with an appropriate record length.

```
open (unit=11, file="xyz.txt", status="old", action="read", &
                          access="direct", recl=80)
```

A file is connected to unit 11 for direct access. The file name is processor-dependent. The file exists already and becomes conected for input only. All records are unformatted (by default) and have the same length of 80 (characters).

Record number

Each record is unambiguously identified by a natural number (= positive integer value), the **record number**. The record number of a particular record is specified by the programmer and established when the record is written. This record number cannot be changed. A record with a particular record number cannot be deleted, but it can be rewritten.

As long as a file is opened for direct access, READ and WRITE statements for the connected unit must include the REC= specifier (with a record number). A record of a file which is opened for direct access may be read only if it has been written since the file was created.

Record length

The record length of a direct access file must be specified in the same OPEN statement which specifies the direct access attribute. All records of a direct access file have the same length. This length is independent of the length of the output list when the records are written. If the length of the output list in a formatted direct WRITE statement is less than the record length, the rest of the record is filled with blanks. But if the length of the output list in an unformatted direct WRITE statement is less than the record length, the rest of the record is undefined. Note that the length of the output list in a direct WRITE statement must never exceed the record length specified in the OPEN statement for the connected unit.

```
character (len=8), dimension (9) :: text
open (unit=7, file="abc.txt", status="new", action="write", &
                          access="direct", recl=72)
:
```

```
do i=10,50,5
  write (unit=7, fmt="(9a)", rec= i) text
  ⋮
enddo
```

The records 10, 15, 20, ..., 45, and 50 are written. These records may be read immediately after they have been written. There are no records written for any other record numbers; therefore, other records cannot be read.

11.3.3   Form of a File

The records of a file must be either *all* formatted or *all* unformatted, except that the last record may be an endfile record. Therefore, files may be classified as *formatted files* and *unformatted files*. The set of allowed forms for a file is processor-dependent.

11.3.4   File Position

A file which is connected to a unit has a **position**. The position may become *indeterminate.*

11

The **initial point** of a file is the position just before the first record. The **terminal point** is the position just after the last record.

If the file is positioned *within* a record, this record is the **current record**; otherwise there is no current record.

The record before the current record is the **preceding record**. If the file is positioned between two records, the first of these records is the preceding record.

The record after the current record is the **next record**. If the file is positioned between two records, the second of these records is the next record. If the file is positioned at its initial point, the first record is the next record.

File position prior to data transfer

Sequential input: If the file is positioned *within* a record, the file position is not changed. Otherwise, the file is positioned at the beginning of the next record. And this record becomes the current record.

Sequential output: If the file is positioned *within* a record, the file position is not changed. Otherwise, a new record is created. This record becomes the last and current record of the file, and the file is positioned at the beginning of this new record.

Direct input/output: The file is positioned at the beginning of the record with the specified record number. This record becomes the current record.

File position after data transfer

If an error condition occurs, the file position is indeterminate.

Nonadvancing input: If no error condition, no endfile condition, but an end-of-record condition occurred, the file is positioned after the record just read.

Nonadvancing input/output: If no error condition, no endfile condition, and no end-of-record condition occurred, the file position is not changed.

Advancing input/output: If no error condition and no endfile condition occurred, the file is positioned after the record just read or written.

Sequential input: If no error condition occurred while the endfile record is read, but if the endfile condition occurred, the file is positioned after the endfile record. At this position no READ, WRITE, and PRINT statements must be executed for the connected unit. However, the file may be repositioned (before any data transfer) by the execution of a BACKSPACE or REWIND statement.

11.4   Units

If an input/output statement refers to an external file, a *unit* must be specified which is *connected* to the file. These units are the F specific processor-independent means of referring to a file, though the set of allowed *unit numbers* is processor-dependent.

A unit may be **connected** to an external file; this operation is described as the *opening* of the file.

Data transfer statements and positioning statements may be executed only for such external files which are connected to a unit. OPEN, CLOSE, and INQUIRE statements may also be executed for external files which are *not* connected to a unit.

A unit may be connected to an external file either by the execution of an OPEN statement, this is described as *explicit opening*, or they may be *preconnected*. If they are preconnected, the connection is automatically established when the program is executed; this is described as *implicit opening*.

A unit must not be connected to more than one file at the same time. And a file must not be connected to more than one unit at the same time. A unit and a file may be disconnected by the execution of a CLOSE statement. After a unit has been disconnected, it may be connected to the same file or to another external file, and the file may be connected to the same unit or to another unit.

11.5   Preconnected Units and Predefined Files

A unit and an external file are **preconnected** if they are automatically connected at the beginning of the execution of the program. This means that a preconnected unit may be used in all input/output statements without prior execution of an OPEN statement for this unit and this file.

In the operating system environment which controls the execution of the program, there may be particular files with processor-dependent function such as *device files* and *standard files*.

Usually, device files and standard files are connected to *standard units* for sequential formatted input/output; for example, F processors may use unit number 5 for the standard input (unit) and unit number 6 for the standard output (unit).

The **standard input unit** (abbreviated: standard input) is the unit which may be specified in a READ statement (with a control information list) as the unit asterisk * or which is implicitly used when a READ statement without a control information list is executed.

The **standard output unit** (abbreviated: standard output) is the unit which may be specified in a WRITE statement (with a control information list) as the unit asterisk * or which is implicitly used when a PRINT statement is executed.

It is processor-dependent which devices are associated with standard input and standard output. In an interactive environment, the standard input device is usually the keyboard and the standard output device is the screen. And in a batch environment, the standard input is usually connected to a predefined file, and the standard output is associated with the system printer.

Often, external files are preconnected for sequential input/output. Some F processors may allow a unit to be connected to a file outside the F program in the operating system environment before the execution of the program. In these cases, possibly no explicit opening of the file is necessary. If the file has no name, the F processor may generate a processor-dependent file name.

11.6   Input/Output Statements

The input/output statements for data transfer are PRINT, READ, and WRITE. The input/output statements for inquiring about, specifying, and changing of properties of files and units are CLOSE, INQUIRE, and OPEN. The input/output statements for the positioning of external files are BACKSPACE, ENDFILE, and REWIND.

```
print "(a,i4)", " test variable mass", mass
read (unit=5, fmt=*) a, b, c
write (unit=9) n, x(1:100,5), x/n
close (unit=7, iostat= ivar, status="keep")
inquire (unit=14, position= pos)
open (unit=77, action="readwrite", status="scratch")
backspace (unit=10)
endfile (unit=77, iostat= ios)
rewind (unit=13)
```

Side effects are not allowed during the execution of an input/output statement. That is, a function reference must not appear anywhere in an input/output statement if the invocation of the function causes another input/output statement to be executed.

11.6.1   Input/Output Specifiers

Each input/output statement includes one or more **input/output specifiers**. These specifiers are used to specify the unit, the file name, the form of the input/output, the access method, the error handling, and so on (see the last example). In the following, those input/output specifiers are described that may appear in the *control information lists* of READ and WRITE statements and, partly, in other input/output statements.

Source or destination of the data to be transferred:	**UNIT = u**
Kind of editing of the external data:	**FMT = format**
Identification of a record for direct io:	**REC = record_number**
Advancing or nonadvancing input/output:	**ADVANCE = yes_or_no**
Output of input/output status information:	**IOSTAT = status_variable**
Output of number of read characters:	**SIZE = number_of_chars**

11.6.1.1   UNIT= Specifier

The UNIT= specifier is used to specify a unit (referring to an external file) or to specify an internal file.

UNIT = u

The **u** may be:

– A nonnegative scalar integer expression; its result is the **unit number**;
– An asterisk * specifying a standard unit which is preconnected for formatted sequential input or for sequential formatted output;
– Or a variable of type character identifying an internal file.

Unit number: The unit number is global to the program; that is, a particular unit number identifies in different program units of a program the same unit for the reference to an external file. The set of allowed unit numbers is processor-dependent.

11.6.1.2   FMT= Specifier

The FMT= specifier is used to specify a format or list-directed formatting for a formatted input/output statement.

FMT = format

The **format** may be:
– A variable of type character defined with a format specification;
– Another character expression which evaluates to a valid format specification;
– Or an asterisk * specifying list-directed formatting.

11.6.1.3   REC= Specifier

The REC= specifier is used in a direct input/ouput statement to specify the record number of the record to be transferred.

REC = record_number

The **record_number** must be a scalar positive integer expression.

```
write (unit=16, rec= i+25) sort, location, year
```

11.6.1.4   ADVANCE= Specifier

The ADVANCE= specifier is used to specify whether it is an advancing or nonadvancing READ or WRITE statement.

ADVANCE = yes_or_no

yes_or_no is a scalar default character expression which must evaluate to "yes" if it is an advancing (that is, a 'normal' record oriented sequential formatted) input/output statement or which must evaluate to "no" if it is a nonadvancing input/output statement. If the ADVANCE= specifier is absent, the default is ADVANCE="yes".

The result of the evaluation of the character expression may include trailing blanks, but they are not significant.

11.6.1.5   End-of-Record Condition

The **end-of-record condition** occurs when a nonadvancing READ statement attempts to transfer data from a position beyond the end of the record. The end-of-record condition is no error condition. When an end-of-record condition occurs during the execution of a nonadvancing READ statement:

- The record is padded with blanks if the record contains fewer characters than are needed according to the input list and the associated format specification;

- The execution of the READ statement terminates;

- The file is positioned after the read record; and

- If the IOSTAT= specifier and the SIZE= specifier are present, the SIZE= specifier becomes defined with the number of characters which have been transferred under format control. The input list items which have been transferred already do *not* become undefined when the end-of-record condition occurs. And the remaining input list items which have not been transferred retain their definition status.

If the end-of-record condition occurs during the execution of a READ statement without an IOSTAT= specifier, program execution is terminated.

11.6.1.6   IOSTAT= Specifier

The IOSTAT= specifier is used to specify a scalar variable of type default integer for input/output status information. These status informations are partly processor-dependent.

IOSTAT = status_variable

After execution of an input/output statement that includes the IOSTAT= specifier (but before the execution of the next input/output statement), the specified status variable becomes defined as follows:

0 (zero) : when no error condition, no end-of-file condition, and no end-of-record condition occurs.

> 0   : when an error condition occurs.

< 0   : when a sequential READ causes the occurrence of the end-of-file condition but no error condition.

< 0   : (with a negative value different from the end-of-file value) when a nonadvancing READ causes the occurrence of the end-of-record condition but no error condition.

These possible values of the status variable are processor-dependent except the value zero.

When during the execution of an input/output statement that includes the IOSTAT= specifier the end-of-file condition, the end-of-record-condition, or an error condition occurs, program execution is *not* terminated but continues with the next executable statement immediately after the input/output statement.

```
integer :: io_status
read (unit=12, fmt="(f10.2)", iostat=io_status) measurement
if (io_status < 0) then
  call eof()                ! end-of-file condition on unit 12
elseif (io_status > 0) then
  call error(io_status)     ! error condition on unit 12
endif
```

11.6.1.7 Error Conditions

The set of error conditions that may occur during the execution of an input/output statement is processor-dependent. When an error condition occurs during the execution of an input/output statement, the execution of the statement terminates, all items in the input list become undefined, and the position of the file becomes indeterminate.

When an error condition occurs during the execution of an input/output statement that does not include the IOSTAT= specifier, program execution is terminated.

11.6.1.8 End-of-File Condition

The **end-of-file condition** occurs when a sequential READ statement attempts to transfer data from a position beyond the end of the file. This may happen for an external file and for an internal file. In the case of a sequential external file, the end-of-file condition occurs when the READ statement encounters an endfile record.

When the end-of-file condition occurs, the execution of the READ statement terminates and all items in the input list become undefined.

When the end-of-file condition occurs during the execution of a READ statement that does not include the IOSTAT= specifier, program execution is terminated.

11.6.1.9 SIZE= Specifier

The SIZE= specifier is used in a nonadvancing READ statement to specify a scalar integer variable which becomes defined with the count of the characters transferred by the execution of this READ statement.

```
integer :: chars
read (unit=37, fmt="(2a)", advance="no", size= chars) syl, letter
if (chars > 4) then
  call word_division()
endif
```

When the nonadvancing READ statement terminates, the SIZE= specifier becomes defined with the count of the characters which have been transferred under format control, that is, under control of data edit descriptors. Blank characters which have been transferred as *padding characters* are not counted.

11.6.2   Input/Output Lists

An **input/output list** specifies the entities whose values are transferred by READ, WRITE, or PRINT statements. The list items must be specified in the same order from the left to the right in which they are to be processed. An input/output list may specify any number of list items; it may be empty.

[**io_list_item** [, **io_list_item**]...]

An *input list* (see below) must specify only input list items and an *output list* (see below) must specify only output list items.

Noneffective list items: A zero-size array or an array section of size zero may be specified as an input/output list item but they are **not effective**; that is, they are ignored as list items during the data transfer.

Arrays: If an input/output list specifies an array, it is treated as though all elements (if any) were specified in array element order. If the input/output list item is an allocatable array, it must be currently allocated.

Derived type entities: If a derived type entity appears as an input/output list item in a formatted input/output statement, it is treated as though all of its components were specified in the same order as in the type definition.

If a derived type entity appears as an input/output list item in an unformatted input/output statement, it is treated as a single value in processor-dependent form.

A derived type entity may appear as an input/output list item only if all components ultimately contained within the entity are accessible within the scoping unit containing the input/output statement.

A derived type entity which ultimately contains a pointer component must not appear as an input/output list item.

Pointers: If an input/output list item is a pointer, data are transferred between the associated target and the file.

Processing of list items: The input/output list is processed from the left to the right; that is, the processing of the next input/output list item does not begin until the value(s) to or from the current input/output list item are transferred. The significance of this rule is demonstrated by the following example:

```
read (unit= n) n, a(n)
```

The old value of n identifies the unit. When the data transfer for this unit begins, the new value for n is transferred. Then the position of a(n) is determined using this new value of n before the value for a(n) is transferred.

Input list

The input/output list in a READ statement must be an *input list*. An **input list item** must be a variable.

```
read (unit=8, fmt="(4f8.2)") a, b, c, d
read (unit=11, fmt=*) a, e(i), f(5, 10), g
read "(2f4.0,i4)", x, y(5*j+3, 10*k-7), l
read (unit=13) ff, ge(k, l, m+3)
```

Empty input list: If the input list is empty or if all list items are noneffective, one or more records are read without transferring data. An unformatted or list-directed READ statement skips exactly one record. For a formatted (but not list-directed) READ statement, the number of skipped records is one plus the number of slash edit descriptors in the format specification.

There are the following restrictions on input list items:

- A variable enclosed in parentheses must not appear as an input list item.

- An input list item in a formatted READ statement must not contain any part of the format specification.

- If an input list item is a pointer, this pointer must be currently associated with a definable target when the READ statement is executed.

- If an input list item is an array section, no array element must affect the value of any expression in the section-subscript list of the input list item and no array element must appear more than once within this array section.

```
integer, dimension (100) ::  a
read (unit=5, fmt=*) a(a)                  ! not allowed
read (unit=5, fmt=*) a(a(1) : a(10))       ! not allowed
```

Output list

The input/output list in a WRITE or PRINT statement must be an *output list*. An **output list item** may be an expression.

```
write (unit=6, fmt=*) 4, g(i, k, k+2), a, p
print "(6es15.4)", t, v, r, r*sin(v), (pi), pi
```

There are the following restrictions on output list items:

- An output list item must be defined before the data transfer begins.

- If an output list item is a pointer, this pointer must be currently associated with a target when the WRITE or PRINT statement is executed.

Empty record: If the output list in an unformatted WRITE statement or in a list-directed WRITE or PRINT statement is empty or if all list items are noneffective, one (empty) record of length zero is written. In the case of a formatted (but not list-directed) WRITE or PRINT statement, the number of written records depends on the edit descriptors in the format specification. The record is empty if the output list is empty or if all list items are noneffective, and if, in addition, the form of the format specification is ( ). An empty record does not contain any data, but it is counted when a READ statement or a BACKSPACE statement is executed.

11.6.3 Data Transfer Statements

The READ statement is the data transfer *input* statement. The WRITE statement and the PRINT statement are the data transfer *output* statements. These three statements are used to perform all kinds of data transfer, as there are sequential, direct, formatted, unformatted, list-directed, nonadvancing, and internal input/output.

READ (control_information_list) [input_list]

READ format [, input_list]

WRITE (control_information_list) [output_list]

PRINT format [, output_list]

The READ and PRINT statements *without* a control information list are *sequential formatted* data transfer statements. If **format** is an asterisk *, they are *list-directed* data transfer statements. In any case, the standard units are used. Because such a statement cannot include an IOSTAT= specifier, the occurrence of an error condition, end-of-file condition, or end-of-record condition causes the program to be terminated.

The data transfer statements *with* a control information list may be used to perform input/output from and to standard units (and standard files) and to perform input/output from and to external files and internal files.

The control information list in a READ or WRITE statement *must* include at least the UNIT= specifier. The appearance or nonappearance of the other input/output specifiers determines the kind of data transfer, the access method, the kind of error handling, the kind of end-of-record handling, and the kind of end-of-file handling to be performed by the READ or WRITE statement. The input/output specifiers may appear in any order within the control information list of a data transfer statement.

If the control information list includes the FMT= specifier, the statement is a *formatted* data transfer statement; otherwise it is an *unformatted* data transfer statement. If the FMT= specifier **format** is an asterisk *, the statement is a *list-directed* data transfer statement (because formatting is list-directed).

If the control information list includes a REC= specifier, the statement is a *direct* data transfer statement; otherwise it is a *sequential* data transfer statement. If the REC= specifier is present, the FMT= specifier **format** must not be an asterisk *.

If ADVANCE="no" is specified, it is a *nonadvancing* data transfer statement. This kind of data transfer reads or writes a record (character-wise) continuously such that one record may be processed by several READ or WRITE statements.

11

If each data transfer statement includes either no ADVANCE= specifier or ADVANCE="yes", then the READ, WRITE, and PRINT statements, possibly except the first one, start transferring data at the beginning of each record. This also is true if the last READ statement did not transfer all data elements of the record. In this case, the remaining data elements are skipped if the file is opened for sequential access. Such a kind of sequential formatted input/output is called *advancing input/output*.

Normally, a data transfer statement reads, writes, or skips at most one record. Only a formatted input/output statement may process more than one record.

Data are transferred between the records in a file and the items in an input/output list. If an input/output list item is a subobject, its current position is determined immediately before the transfer of this subobject begins.

The execution of a data transfer statement terminates:

- When format control encounters a data edit descriptor for which there is no effective input/output list item left;

- When an unformatted or list-directed data transfer has processed all items in the input/output list;

- When an error condition occurs;

- When the end-of-record condition or the end-of-file condition occurs; or

- When a slash "/" is encountered as a value separator in the record being read by a list-directed READ statement.

11.6.3.1 Formatted Input/Output

During formatted data transfer, data are transferred with editing, because, normally, their internal and external representations are different. There are special kinds of formatted input/output such as list-directed, nonadvancing, and internal input/output.

A formatted WRITE or PRINT statement writes data in character-oriented form to the file, and the formatted READ statement reads data in character-oriented form from the file. This character-oriented data might be read also by a human reader. The conversion (or editing) between the internal form and the external character-oriented form may be very expensive. This kind of input/output is usefull only if the data in the file are to be read by a human reader or are to be interchanged with another computer model.

READ (control_information_list) [**input_list**]

READ format [**, input_list**]

WRITE (control_information_list) [**output_list**]

PRINT format [**, output_list**]

The following input/output specifiers must be or may be present for formatted input/output: UNIT = **u**, ,FMT = **format**, REC = **record_number**, ADVANCE = **yes_or_no**, IOSTAT = **status_variable**, and SIZE = **number_of_chars**. The UNIT= specifier and the FMT= specifier *must* be present; the other specifiers *may* be present.

Statements without a control information list

The READ and PRINT statements *without* a control information list are *sequential formatted* data transfer statements. They use the standard units. If **format** is an asterisk *, they are *list-directed* data transfer statements.

Statements with a control information list

The kind of formatted input/output is given by the presence of suitable input/output specifiers. The following table shows which specifier(s) *must* be present, which *may* be present, and which must *not* be present:

Formatted io:	UNIT	FMT	REC	ADVANCE	IOSTAT	SIZE
list-directed	must	asterisk	not	not	may	not
nonadvancing	2)	must	not	"no"	may	5)
internal	3)	must	not	not	may	not
direct	1)	4)	must	not	may	not
advancing	must	must	not	6)	may	not
Notes:	1) must; but no character variable					
	2) must; but no asterisk * and no character variable					
	3) character variable					
	4) must; but no asterisk *					
	5) may; but only in READ statement					
	6) may; but only with "yes"					

```
character (len=80) :: text
character (len=8), dimension (10) :: word
real, dimension (10)   :: is
real, dimension (100) :: y
⋮
read (unit=10, fmt="(3f7.2)") sort, area, qty  ! sequ. form. input
print *, a, b, c                               ! list-dir. output
write (unit=23, fmt="(2i6)", advance="no") l, b  ! nonadv. output
read (unit= text, fmt="(10a)") word            ! internal input
write (unit=22, fmt="(110f7.2)", rec=5) is, y  ! dir. form. output
```

If the ADVANCE= specifier is absent or if ADVANCE="yes" is present, a formatted data transfer statement processes one or more records and transfers data (according to a format specification) between the items of the input/output list and the specified unit, which is connected to a file. If ADVANCE="no" is present, a single record may be processed character-wise by the execution of more than one data transfer statement.

If a formatted direct access data transfer statement processes more than one record, the record number is increased by one as each succeeding record is read or written.

The input/output list items must match the corresponding data edit descriptors. If a file is opened for unformatted input/output, formatted input/output is not allowed.

Formatted input (*without* list-directed formatting)

An input list item (or an associated variable) must not contain any part of the format specification.

If the input list and the corresponding format specification attempt to read more data from the record than the record contains, padding characters (i. e., blanks) are supplied for the missing characters.

When a sequential READ statement encounters the endfile record in an external file or when it attempts to read beyond the end of an internal file, program execution is terminated if the IOSTAT= specifier is not present in the READ statement.

Formatted output (*without* list-directed formatting)

If the characters specified by the output list and the format of a direct or internal WRITE statement do not fill the record, the rest of the record is filled with blanks.

In the case of an external file, the output list must match the format specification such that the output statement does not attempt to write more characters in the record than have been specified by the RECL= specifier in the OPEN statement.

And an internal WRITE statement must not attempt to write more characters in the record than the record length of the internal file.

```
real, dimension (10)  :: tree
real, dimension (100) :: y
:
write (unit=22, fmt="(10i5,100f7.2)", rec=152) tree, y
```

The WRITE statement writes (with direct access and controlled by the specified format) 10 values of array tree and 100 values of array y into the record with record number 152.

11.6.3.2 Unformatted Input/Output

An unformatted data transfer statement does not need a format specification, because the data are transferred in their internal binary form from and to external files. That is, there is no difference between the internal and the external representations of the data.

Because unformatted input/output is simpler and cheaper than formatted input/output, this kind of data transfer is usefull for intermediate or temporary files. Usually, unformatted files created by one computer cannot be processed by another computer model.

READ (control_information_list) [input_list]

WRITE (control_information_list) [output_list]

The following input/output specifiers must be or may be present for unformatted input/output: UNIT = **u**, REC = **record_number**, and IOSTAT = **status_variable**. The UNIT= specifier *must* be present; the other specifiers *may* be present.

The kind of unformatted input/output is given by the presence of suitable input/output specifiers. The following table shows which specifier(s) *must* be present, which *may* be present, and which must *not* be present:

Unformatted io:	UNIT	FMT	REC	ADVANCE	IOSTAT	SIZE
sequential	1)	not	not	not	may	not
direct	1)	not	must	not	may	not
Notes:	1) must; but no asterisk * and no character variable					

```
real, dimension (100)    :: x
real, dimension (20, 10) :: z
⋮
read (unit=7, iostat= is) n, x(1:n)  ! sequential unformatted input
write (unit=47, rec=12) fe, z        ! direct unformatted output
```

If a file is opened for formatted input/output, unformatted data transfer statements are not allowed for that file.

11

Unformatted input

An unformatted READ statement may read only such records which have been written before by unformatted WRITE statements.

Each value in the input record must be of the same type as the corresponding input list item, except two real values may correspond to one complex input list item, and one complex value may correspond to two real input list items. The kind type parameter or the character length of the value must agree with that of the corresponding list item.

The number of values required by the input list must be less than or equal to the number of values in the record. If the input list requires fewer values than the record contains, the rest of the values is ignored.

```
real, dimension (100) :: x
read (unit=12, iostat= ista) n, x(1:n)
```

The unformatted READ statement reads with sequential access the file which is connected to unit 12. A value is read in the scalar integer variable n. Then n values are transferred into the first n elements of array x. After execution of the READ statement, the variable ista becomes defined with a status value.

Unformatted output

A direct access WRITE statement must not transfer more values into the record than have been specified by the RECL= specifier in the OPEN statement.

If the file is opened for sequential access and if the maximum record length is specified by the RECL= specifier in the OPEN statement, then a WRITE statement must not attempt to transfer more values than have been specified by the RECL= specifier.

If an unformatted direct access WRITE statement writes fewer values than the record length, the rest of the values in the record is undefined. If the output list is absent, an empty record is written.

```
real :: fe
real, dimension (20, 10) :: z
:
write (unit=47, rec=12) fe, z
```

The unformatted WRITE statement writes (with direct access to the file which is connected to unit 47) the value of the variable fe and the 200 values of array z into the record with the record number 12. When an error condition occurs, program execution is terminated.

11.6.3.3   List-Directed Input/Output

List-directed input/output is a special kind of sequential formatted input/output from and to external files or internal files. The correct term is "formatted input/output with list-directed formatting". Though formatted records are processed, the F programmer need not and cannot specify a format, because, depending on the type of the input/output list items, certain internal predefined edit descriptors are used.

READ (control_information_list) [input_list]

READ * [, input_list]

WRITE (control_information_list) [output_list]

PRINT * [, output_list]

The following input/output specifiers must be or may be present for list-directed input/output: UNIT = u, FMT = *, and IOSTAT = **status_variable**.   The UNIT= specifier and the FMT= specifier *must* be present; the IOSTAT= specifier *may* be present.

Statements without a control information list

The READ statement and PRINT statement *without* a control information list use the standard units.

Statements with a control information list

The kind of list-directed input/output is given by the presence of suitable input/output specifiers. The following table shows which specifier(s) *must* be present, which *may* be present, and which must *not* be present:

List-directed io:	UNIT	FMT	REC	ADVANCE	IOSTAT	SIZE
external	1)	asterisk	not	not	may	not
internal	2)	asterisk	not	not	may	not
Notes:	1) must; but no character variable					
	2) character variable					

```
character (len=40) :: parents
read (unit=*, fmt=*) current, field, voltage
read *, current, field, voltage
write (unit=*, fmt=*, iostat= ivar) alpha, beta, gamma
print *, alpha, beta, gamma
read (unit= parents, fmt=*) father, mother
```

11

List-directed input

The order of the data in the input record may be almost arbitrary and is not subject to any specifications (like edit descriptors in a format specification). An input record contains a sequence of values, which are separated by *value separators*.

A **value separator** is one of the following:

- One or more contiguous blanks, except one of the two values is a *null value* (see below) without a repeat factor. A sequence of two or more blanks is treated as a single blank;

- A comma optionally enclosed in one or more blanks; or

- A slash optionally enclosed in one or more blanks.

A list-directed READ statement may read more than one record. The end of a record has the same effect as a blank character, unless a character value is continued to the next record. That is, if more values are to be transferred than fit in the record, the rest of the values may be continued to the next record.

Normally, a record must contain complete values, but in the case of a complex value or a character value, the end of the record may appear within the value.

Blanks are not interpreted as zeros. Because blanks are used as value separators, they must not appear within values, except significant blanks in character values and insignificant blanks in complex values.

There are some rules governing the form of the external representation of the input values. Generally (with few exceptions), input values which can be read with explicit formatting can also be read with list-directed formatting. The form of an input value must be acceptable by the next effective input list item.

A value in the input record is either an *integer value*, a *real value*, a *complex value*, a *logical value*, a *character value*, a *null value*, or such a value preceded by a repeat factor (see below):

Integer value: An integer value must have the same form as an integer literal constant without an explicitly specified kind type parameter. The corresponding effective input list item must be of type integer.

Real value: A real value may have the same form as a real literal constant without an explicitly specified kind type parameter. It also may have the same form as an integer value; then the input value is interpreted as the significand of a real value with no fractional digits. The corresponding effective input list item must be of type real.

Complex value: A complex value must have the same form as a complex literal constant. Precisely, it consists of two real values, which are separated by a comma and which are enclosed in parentheses. The first value is the real part and the second value is the imaginary part of the complex value. The parentheses are not interpreted as separators. Each of the two values may be enclosed in blanks. The end of the record may occur between the real part and the comma or between the comma and the imaginary part. The corresponding effective input list item must be of type complex.

Logical value: A logical value may have the same form as a logical literal constant without an explicitly specified kind type parameter. Allowed are: an optional point, followed by the letter "t" or "f", followed by any characters except slashes, blanks, equals, and commas. The corresponding effective input list item must be of type logical.

Character value: A character value may have the same form as a character literal constant. If the quote-delimited character value includes a doubled quotation mark, these must not be separated by blanks or by the end of the record. Such a character value may contain blanks, commas, and slashes.

Character values may be continued from the end of one record to the beginning of the next record. If the character value does not have the same length as the input list item, the effect of the data transfer is as though the character value were assigned to the list item in a character intrinsic assignment statement.

The delimiting quotation marks are not required

1. if the character value does not contain a blank, a comma, or a slash; and
2. if the character value is not continued to the next record(s); and
3. if the first nonblank character is neither an apostrophe nor a quotation mark; and
4. if the leading characters are no digits followed by an asterisk *; and
5. if the character value contains at least one character.

In this case, the character value is terminated by the first blank, comma, slash, or end of record; and quotation marks within the character value are *not* to be doubled.

Null value: A null value is represented either by two consecutive value separators without intervening blanks, or by no characters appearing before the first value separator in the first record read by a list-directed READ statement, or by the form **n*** (see below).

The rest of a record after the last comma or slash, with or without blanks, is not interpreted as a null value. A null value has no effect on the definition status or the value of the corresponding effective input list item. A null value must not be used as the imaginary part or as the real part of a complex value, but a single null value may be used as an entire complex value. The effective input list item corresponding to a null value may be of any type.

Other possibilities of forming an input record:

Repeat factor: The input values may specify a *repeat factor* **n**, which must be an unsigned integer literal constant without an explicitly specified kind type parameter. The following forms are allowed:

n*   is the repetition of the null value. This form is the same as **n** successive appearances of the null value.

n*c   is the same as **n** appearances of the integer, real, complex, logical, or character value **c**. Embedded blanks are not allowed except where permitted in **c**. **c** has implicitly the same kind type parameter value as the corresponding effective input list item.

Slash: When during the execution of a list-directed READ statement a *slash* as a value separator is encountered, the execution of this READ statement terminates after the transfer of the previous value. Any remaining characters in the input record are ignored. The effect is as though null values were transferred to those input list items (if any) which have not been processed.

```
do
  read (unit=*, fmt=*, iostat= ios) l, m
  if (ios < 0) then
    exit
  endif
  write (unit=9, fmt="(i5,i6)") l, m
enddo
```

Input records: 5 70

                     ,   1

                     5*2

                     , ,

                     /   7

                     12345678901   ⟵ record position

Output records: 5   70

                         5   1

                         2   2

                         2   2

                         2   2

                     12345678901   ⟵ record position

List-directed output

A list-directed WRITE or PRINT statement writes the values of the output list items in principle in a form to the output record that corresponds to the form of the input values. The F processor uses its own (sensible) internal edit descriptors.

Character values are not delimited by quotation marks.

The programmer has nearly no influence on the form of the external representation of the output data and on the form of the output record. Therefore, this kind of output is usefull mainly during the test phase of the program development cycle.

11.6.3.4   Internal Input/Output

Internal input/output is a special kind of sequential formatted input/output either with explicit formatting or with list-directed formatting. Internal input/ output supports the transfer of data from one place in internal storage to another place in internal storage without involvement of an input/output device or an external file.

During internal data transfer, the representation of the data may be changed. Depending on the kind of the internal input/output, the kind of data conversion and the "external" form of the data is explicitly specified by a format specification or is implicitly specified by the F processor. An *internal file* is used instead of an external file.

READ (control_information_list) [input_list]

WRITE (control_information_list) [output_list]

The following input/output specifiers must be or may be present for internal input/output : UNIT = **u**, FMT = **format**, and IOSTAT = **status_variable**. The UNIT= specifier and the FMT= specifier *must* be present; the IOSTAT= specifier *may* be present.

The kind of internal input/output is given by the presence of suitable input/output specifiers. The following table shows which specifier(s) *must* be present, which *may* be present, and which must *not* be present:

Internal io:	UNIT	FMT	REC	ADVANCE	IOSTAT	SIZE
formatted	1)	must	not	not	may	not
list-directed	1)	asterisk	not	not	may	not
Notes:	1) character variable					

```
character (len=40) :: 140
character (len=80) :: ad80
read (unit= 140, fmt="(i8,tr6x,f8.2,es10.4)") max, r_min, sel
write (unit= ad80, fmt=*, iostat= k_stat) value, "no instr", year
```

Internal file

An **internal file** is a variable of type character, which is specified by the UNIT= specifier **u**. If **u** is an array section, it must not have a vector subscript.

Internal files must not and cannot be opened explicitly or implicitly. If an internal file is an allocatable array which is currently not allocated, or if the internal file is a subobject of such an array, or if the internal file is a pointer which is currently not associated with a target, then the internal file is inaccessible and must not be involved in data transfers.

Record: A record of an internal file is a *scalar* character object. If the internal file is a scalar variable, the internal file consists of a single record with a record length which is given by the length of the character variable.

If the internal file is an *array* (that is, a whole array or an array section), the internal file is treated as a sequence of array elements in array element order, and each array element is a record of the internal file. In this case, all records of the internal file have the same record length, which is the character length of the single array elements.

A record of an internal file must be read only if it is defined.

Prior to data transfer, an internal file is always positioned before the first character position of its first record.

Input/output statements: File positioning statements (BACKSPACE, REWIND, ENDFILE) and file status statements (OPEN, CLOSE, INQUIRE) are not applicable to internal files.

Input/output list: An input/output list item must neither be a part of the internal file nor be associated with the internal file.

Internal input

Beginning with the first character position of the internal file, data are read and transferred to the input list items under control of the supplied format specification or under control of list-directed formatting. During internal input, the data may be converted from the external representation to an internal representation.

Blanks: During list-directed formatting, blanks are normally treated as value separators. Under control of an explicit format specification, single blanks are ignored and an input field consisting of all blanks is interpreted as the value zero.

Padding characters: If an internal READ attempts to read more characters than the record contains, blank characters are supplied for the missing characters.

```
character (len=40), save :: &
                   line = "15.00 $ for  750 g,   20.00 $ per kg"
real :: price, kg
integer :: weight
read (unit=line, fmt="(f5.2,tr8,i4,tr6,f5.2)") price, weight, kg
```

The internal file line contains one record with record length 40. After execution of the READ statement, the variables price, weight, and kg are defined such:

    price = 15.0   weight = 750   kg = 20.0

Internal output

The values of the output list items are written to the internal file under control of the specified format specification or under control of list-directed formatting.

During internal data transfer, the data may be converted from the internal representation to an external representation. The data transfer begins always at the first character position of the file, even if the file has more than one record.

If the number of output characters is less than the record length, the rest of the record is filled with blanks. The number of output characters must not exceed the record length.

During list-directed formatting, character values are written without delimiting quotation marks.

The format specification for the internal WRITE statement must not be contained in the internal file or in an associated data object.

```
character (len=3), dimension (8), save :: digits = "***"
character (len=3), parameter :: a = "uno", b = "due", c = "tre"
write (unit=digits, fmt="(a/)") a, b, c
```

The internal file digits contains 8 records with record length 3. After execution of the WRITE statement, these 8 records are defined as follows:

bafw(1) = "uno"   bafw(2) = "␣␣␣"   bafw(3) = "due"   bafw(4) = "␣␣␣"
bafw(5) = "tre"   bafw(6) = "␣␣␣"   bafw(7) = "***"   bafw(8) = "***"

When format control for an internal (or direct) WRITE statement encounters the characters "/)" in the format specification, a new record is defined (created) which is filled with blanks.

11.6.3.5   Nonadvancing Input/Output

Nonadvancing input/output is a special kind of sequential formatted input/output from and to external files with explicit formatting. List-directed formatting and internal files are not supported. Nonadvancing input/ output transfers the characters of a record (character-wise) continuously such that one record may be processed by several data transfer statements. Nonadvancing input supports processing of records of varying (unknown) lengths.

READ (control_information_list) [input_list]

WRITE (control_information_list) [output_list]

The following input/output specifiers must be or may be present for nonadvancing input/output: UNIT = **u**, FMT = **format**, ADVANCE = **yes_or_no**, IOSTAT = **status_variable**, and SIZE = **number_of_chars**. The UNIT= specifier, the FMT= specifier, and the ADVANCE= specifier *must* be present; the other specifiers *may* be present.

Nonadvancing input/output is given by the presence of suitable input/output specifiers. The following table shows which specifier(s) *must* be present, which *may* be present, and which must *not* be present:

Nonadvancing io:	UNIT	FMT	REC	ADVANCE	IOSTAT	SIZE
formatted	must	1)	not	2)	may	3)
Notes:	1) must; but no asterisk * and no character variable					
	2) must; **yes_or_no** must be "no"					
	3) may; but only in READ statement					

```
read (unit=14, fmt="(a4,i4)", advance="no") b, m ! nonadv. input
write (unit=18, fmt="(i1)", advance="no") n        ! nonadv. output
```

Nonadvancing input/output differs from the "advancing" forms of input/output with regard to the way the file is positioned after the data transfer. After nonadvancing data transfer, the file position is frozen and there is *no* automatic positioning after the record just read or written. If no end-of-record condition, no error condition, and no end-of-file condition occur, the file remains positioned within the current record. Then the next data transfer statement may continue to process the current record at this position.

When a slash edit descriptor is encountered or when format control reverts, the file is positioned in the usual manner as for the advancing forms of input/output.

The T, TL, and TR edit descriptors are processed in the same manner as for advancing input/output, except that the file must not be positioned to the left of the *left tab limit*.

If the file is opened for formatted input/output, advancing and nonadvancing data transfer statements may be executed in any order.

Nonadvancing input

In the following, we assume that neither an error condition nor the end-of-file condition occurs.

If during the execution of a nonadvancing data transfer statement no end-of-record conditon occurs, there is a current record after the data transfer, and the record is positioned within this current record just after the last transferred character. The next READ statement begins reading at this position.

If there is no current record, the next READ statement begins reading at the first character position of the next record.

If during the execution of a nonadvancing READ statement the end-of-record condition occurs, the execution of the READ statement terminates. The already transferred input list items do *not* become undefined. And depending on the absence or presence of the IOSTAT= specifier, program execution is terminated or continued.

A nonadvancing READ statement may include the SIZE= specifier. After the execution of the READ statement, this SIZE= specifier is defined with the count of characters which have been transferred under control of data edit descriptors.

```
character (len=8) :: word
integer :: wl, ios
read (unit=9, fmt="(a)", iostat= ios, size= wl, advance="no") word
```

The nonadvancing READ statement reads a character value from the file which is connected to unit 9. If no end-of-record condition occurs, the variable wl is redefined with the value 8, because exactly 8 characters have been transferred. But if the end-of-record condition occurs, the variable wl is redefined with the number of characters which have been transferred before the READ statement has encountered the end of the record.

Nonadvancing output

In the following, we assume that no error condition occurs.

If during the execution of a nonadvancing data transfer statement no end-of-record conditon occurs, there is a current record after the data transfer, and the record is positioned within this current record just after the last transferred character. The next WRITE statement begins writing at this position.

If there is no current record, the next WRITE statement begins writing at the first character position of the next record.

```
integer, dimension (10) :: nu, m
 ⋮
write (unit=47, fmt="(2i10)", advance="no") nu
write (unit=47, fmt="(2i10)", advance="no") m(1), m(3), m(5)
write (unit=47, fmt="(2i10)", advance="yes") m(7), m(9)
write (unit=47, fmt="(4i10)", advance="no") m(2), m(4), m(6), m(8)
```

The first (nonadvancing) WRITE statement transfers 5 records, each containing 20 characters, to the file which is connected to unit 47. The 5th record is the current record after the execution of this WRITE statement; that is, the file is positioned within this record. The next (nonadvancing) WRITE statement first extends the current record by 20 characters and then writes additional 10 characters in the next record. This 6th record is now the current record. The next (advancing) WRITE statement extends the current record by additional 20 characters and terminates the record. The last (nonadvancing) WRITE statement creates the first characters of the 7th record.

11.6.3.6   Printing

Within Fortran, sequential formatted output to particular (processor-dependent) predefined output devices is called **printing**. The first character of each output record is not written on the output device, but it is used as a *carriage control character* that controls the vertical spacing of the output device. The remaining characters of the record are written beginning at the first character position of the printed *line*.

The F language does not directly support this kind of printing. Since printing may depend on the file, the operating system, and the device, it is processor-dependent. In UNIX environments, vertical spacing of output devices is not controlled by carriage control characters but by ASCII control characters. And the user has to convert Fortran files to appropriate UNIX files before "printing".

For printing in accordance with Fortran, the **carriage control character** must be of type character. It controls the vertical spacing *before* printing:

Character	Vertical spacing before printing
Blank	One line.
+	No advance.
1	To first line of next page.
0	Two lines; i. e., print one line with blanks, then advance one line.

The carriage control character + supports printing of more than one record in one line.

If a formatted record will be written but never be printed, the record need not contain a carriage control character. But if it still contains a carriage control character, this character belongs to the data and is written as the first character of the data.

An empty record is printed as a new line containing only blanks.

The first character of *each* printed record must be a carriage control character. This is necessary even if a single PRINT or WRITE statement prints more than one record when format control reverts or when format control encounters a slash edit descriptor. Carriage control characters may be created by different methods.

```
character (len=*), parameter :: f33 = "("""+Result = """,f10.4,i15)"
character (len=1), save :: v = "+"
print f33, y, n
write (unit=*, fmt="(a,f10.4,i15)") v, x, i
```

11.6.4 File Status Statements

There are three file status statements: OPEN, CLOSE, and INQUIRE. They are used to specify properties (attributes) of external files and units, to modify certain properties of external files and units, and to inquire about properties of external files and units.

11.6.4.1 OPEN Statement

The OPEN statement may be used to **open** an external file, that is, to connect the file to a unit. An existing external file may be connected to a unit, a nonexisting external file may be created and then connected to a unit, an external file may be created which is preconnected to a unit, or particular properties of a connection between an external file and a unit may be specified or modified.

Most of the optional specifiers that may be supplied in an OPEN statement accept scalar character expressions. Their valid results are described in the following. Trailing blanks are allowed but are ignored. Some specifiers will be used with default values if they are not explicitly specified. The UNIT=, ACTION=, and STATUS= specifiers *must* be specified, the other specifiers *may* be specified. Within the source text, the specifiers and their values must be written in lower case. The specifiers may appear in any order.

11

```
OPEN ( UNIT = u, ACTION = action, STATUS = status
         [, IOSTAT = iostat] [, FILE = file] [, ACCESS = access]
         [, FORM = form] [, RECL = recl] [, POSITION = position])
```

u	is a scalar nonnegative integer expression specifying the number of the unit which is to be connected to the file being opened.
action	is a character expression specifying the allowed data transfer actions for the file.

"read"	WRITE, PRINT and ENDFILE are not allowed.
"write"	READ is not allowed.
"readwrite"	All input/output statements are allowed.

status	is a character expression specifying the status of the file being connected to unit **u**.

"old"	The file exists already.
"new"	The file does not yet exist.
"scratch"	The file is to be deleted at the execution of a CLOSE statement for this unit or at program termination.
"replace"	If the file does not exist, it is created. If it exists, it is deleted and a new file is created.

iostat   is a scalar status variable of type default integer which will become defined with a zero value if no error condition occurs during the execution of this file status statement; otherwise it becomes defined with a processor-dependent positive error status value.

file   is a character expression specifying the name of the file being opened. This file becomes connected to unit **u**. The form of the file name is processor-dependent.

access   is a character expression specifying the access method.

   "sequential"   The file is being opened for sequential access.

   "direct"   The file is being opened for direct access.

   Default: **ACCESS = "sequential"**.

form   is a character expression specifying the form of the file.

   "formatted"   The file is being opened for formatted input/output.

   "unformatted"   The file is being opened for unformatted input/ output.

   Default: **FORM = "formatted"**   if access is sequential and
   FORM = "unformatted"   if access is direct.

recl   is a scalar positive integer expression specifying either the record length in a file being connected for direct access or the maximum record length in a file being connected for sequential access.

position   is a character expression specifying the position of the (existing sequential) file immediately after opening.

   "rewind"   The file is positioned at its initial point.

   "append"   The file is positioned at its terminal point (if possible, before the endfile record) such that the file can be extended.

   "asis"   The position of the file is not changed. The position is processor-dependent if the file exists but has not been connected.

   Default: **POSITION = "asis"**

IOSTAT= specifier: When an error condition occurs during the execution of a file status statement that does not includ an IOSTAT= specifier, program execution is terminated.

FILE= specifier: The FILE= specifier must be present if the STATUS= specifier is present and has the value "old", "new", or "replace". The FILE= specifier must be absent if the STATUS= specifier is present and has the value "scratch". If the FILE= specifier is absent, the STATUS= specifier must be present with the value "scratch"; in this case, the unit becomes connected to a processor-dependent file, and the F processor may generate a processor-dependent name for this file.

```
character (len=3) :: head, tail
open (unit=7, file= head//tail, status="new", action="read")
```

The nonexisting file whose name results from the evaluation of the character expression **head//tail** is being connected to unit 7. The file is being opened for sequential access. The file is formatted. WRITE and ENDFILE statements cannot be executed for this file. When an error condition occurs during the execution of this OPEN statement, program execution is terminated.

STATUS= specifier: STATUS="scratch" specifies that the file is a temporary file; the FILE= specifier must be absent and ACTION="readwrite" must be present. If STATUS="old", "new", or "replace" is specified, the FILE= specifier must be present. If a file with STATUS="old" is opened for sequential access, POSITION="rewind" or "append" must be specified. If during the execution of an OPEN statement with STATUS="new" *no* error condition occurs, the file is created and its status becomes "old".

ACCESS= specifier: For an existing file, the specified access method must be included in the set of allowed access methods for the file.

RECL= specifier: For an existing file, the specified record length or maximum record length must be included in the set of allowed record lengths for the file. The RECL= specifier must be present if ACCESS="direct" is specified; otherwise the RECL= specifier may be absent. If the RECL= specifier is absent, the maximum record length is processor-dependent.

POSITION= specifier: Only existing files are affected by the POSITION= specifier. A sequential nonexisting file will be always positioned at its initial point.

Multiple OPENs: An OPEN statement must not be executed for a unit that is connected. After execution of a CLOSE statement for the unit, this unit may be reconnected even with the same file as before.

11.6.4.2 CLOSE Statement

The execution of a CLOSE statement closes an *external* file and disconnects a unit, respectively; that is, the connection between the specified unit and an external file is terminated. Depending on the value of the STATUS= specifier, the file will continue to exist or will be deleted. Within the source text, the specifiers and their values must be written in lower case.

CLOSE (UNIT = u [, IOSTAT = Iostat] [, STATUS = status])

u   is a scalar nonnegative integer expression specifying the number of the unit to be disconnected.

Iostat   (see OPEN statement)

status   is a scalar character expression specifying the disposition of the file to be closed.

      "keep"   The file will *not* be deleted after the execution of the CLOSE statement.

      "delete"   The file will be deleted after the execution of the CLOSE statement.

      Default: **STATUS="delete"** if the file has been opened with STATUS="scratch"; otherwise is **STATUS="keep"**.

UNIT= and IOSTAT= specifiers: (see OPEN statement)

STATUS= specifier: Trailing blanks are allowed but they are ignored. STATUS="keep" must not be specified for an external file that has been opened with STATUS="scratch".

```
close (unit=10, iostat= ios, status="delete")
```

The file which is connected to unit 10 is closed and deleted. If an error condition occurs during the execution of the CLOSE statement, program execution is not terminated.

Program termination: When the execution of a program is terminated in the normal way (that is, if neither an error condition nor another condition occurs), then all files will be automatically closed which are connected in the moment of the termination. The effect is as though a CLOSE statement without a STATUS= specifier were executed for each of these files.

11.6.4.3   INQUIRE Statement

The INQUIRE statement may be used to inquire about the attributes of a particular named external file, about the connection to a particular unit, or about the length of an output list.

There are tree kinds of INQUIRE statements:

- INQUIRE with IOLENGTH= specifier (called an *inquiry by output list*),

- INQUIRE with UNIT= specifier (called an *inquiry by unit*), and

- INQUIRE with FILE= specifier (called an *inquiry by file*).

INQUIRE (IOLENGTH = Iolength) output_list

INQUIRE ($\left\{ \begin{array}{l} \text{UNIT = u} \\ \text{FILE = file} \end{array} \right\}$ **[, IOSTAT = iostat] [, EXIST = exist]**
   [, OPENED = opened] [, NUMBER = number] [, NAMED = named]
   [, NAME = name] [, ACCESS = access] [, SEQUENTIAL = sequential]
   [, DIRECT = direct] [, FORM = form] [, FORMATTED = formatted]
   [, UNFORMATTED = unformatted] [, RECL = recl] [, NEXTREC = nextrec]
   [, POSITION = position] [, ACTION = action] [, READ = read]
   [, WRITE = write] [, READWRITE = readwrite])

The specifiers of the INQUIRE statement may appear in any order. The specifiers returning values must be *scalar* variables of type *default* integer, *default* logical, or character. Character values returned by the INQUIRE statement are produced in upper case by the input/ouput system of the F processor.

u	is a scalar nonnegative integer expression specifying the number of the unit being inquired about.
file	is a scalar character expression specifying the name of the file being inquired about.
iolength	is an integer variable which is assigned the record length that would result from the use of the **output_list** in an unformatted WRITE statement.
iostat	(see OPEN statement)
exist	is a logical variable which is assigned the value *true* if the specified unit exists (when inquired by unit) or if there exists a file with the specified name (when inquired by file); otherwise the variable is assigned the value *false*.
opened	is a logical variable which is assigned the value *true* if the specified unit is connected to a file (when inquired by unit) or if the specified file is connected to a unit (when inquired by file); otherwise the variable is assigned the value *false*.
number	is an integer variable which is assigned the number of the unit that is currently connected to the specified file. If the file is not connected to a unit, the variable is assigned the value -1.
named	is a logical variable which is assigned the value *true* if the file has a name; otherwise the variable is assigned the value *false*.
name	is a character variable which is assigned the name of the file if the file has a name; otherwise the variable becomes undefined.
access	is a character variable which is assigned the value:

"SEQUENTIAL"   if the file is opened for sequential access.

"DIRECT"   if the file is opened for direct access.

"UNDEFINED"   if there is no connection.

11

sequential   is a character variable which is assigned the value:

      "YES"   if sequential access is allowed for the file.

      "NO"   if sequential access is *not* allowed for the file.

      "UNKNOWN" if the F processor is unable to determine whether or not sequential access is allowed.

direct   is a character variable which is assigned the value:

      "YES"   if direct access is allowed for the file.

      "NO"   if direct access is *not* allowed for the file.

      "UNKNOWN" if the F processor is unable to determine whether or not direct access is allowed.

form   is a character variable which is assigned the value:

      "FORMATTED" if the file is opened for formatted input/output.

      "UNFORMATTED"if the file is opened for unformatted input/output.

      "UNDEFINED" if there is no connection.

formatted   is a character variable which is assigned the value:

      "YES"   if formatted is an allowed form for the file.

      "NO"   if formatted is *not* an allowed form for the file.

      "UNKNOWN" if the F processor is unable to determine whether or not formatted is an allowed form for the file.

unformatted   is a character variable which is assigned the value:

      "YES"   if unformatted is an allowed form for the file.

      "NO"   if unformatted is *not* an allowed form for the file.

      "UNKNOWN" if the F processor is unable to determine whether or not unformatted is an allowed form for the file.

recl   is an integer variable which is assigned the record length or the maximum record length (see OPEN statement).

nextrec   is an integer variable which is assigned the value $n+1$ if n is the record number of the last record read from or written to the file connected for direct access. If the file is opened but no records have been transferred since the connection, the integer variable is assigned the value 1. If the file is not opened for direct access or if the position of the file is indeterminate because of a previous error condition, the integer variable becomes undefined.

position   is a character variable which is assigned the value:

    "REWIND"   if the file is connected by an OPEN statement for positioning at its initial point.

    "APPEND"   if the file is connected by an OPEN statement for positioning at its endfile record or at its terminal point.

    "ASIS"   if the file is connected without changing its position.

    "UNDEFINED"   if the file is not opened or if the file is opened for direct access.

action   is a character variable which is assigned the value:

    "READ"   if the file is opened only for input.

    "WRITE"   if the file is opened only for output.

    "READWRITE"   if the file is opened for input and output.

    "UNDEFINED"   if there is no connection.

read   is a character variable which is assigned the value:

    "YES"   if input is an allowed action for the file.

    "NO"   if input is *not* an allowed action for the file.

    "UNKNOWN"   if the F processor is unable to determine whether or not input is an allowed action for the file.

write   is a character value which is assigned the value:

    "YES"   if output is an allowed action for the file.

    "NO"   if output is *not* an allowed action for the file.

    "UNKNOWN"   if the F processor is unable to determine whether or not output is an allowed action for the file.

readwrite   is a character variable which is assigned the value:

    "YES"   if both input and output are allowed actions for the file.

    "NO"   if input and output are *not* allowed actions for the file.

    "UNKNOWN"   if the F processor is unable to determine whether or not both input and output are allowed actions for the file.

When an error condition occurs during the execution of an INQUIRE statement that does not include an IOSTAT= specifier, program execution is terminated.

When an error condition occurs during the execution of an INQUIRE statement, the values of all output specifiers, except the IOSTAT= specifier, become undefined.

The NAME= specifier and the NEXTREC= specifier become defined by the execution of the INQUIRE statement only if there is a connection; otherwise their values become undefined.

The EXIST= specifier and the OPENED= specifier become always defined by the execution of the INQUIRE statement, except when an error condition occurs.

The POSITION= specifier is defined with a processor-dependent value if the file has been repositioned since the connection.

```
logical :: ex
integer :: rl
real, dimension (100) :: z
character (len=15) :: acc, filename, name
inquire (file= filename, exist= ex, access= acc) ! inquiry by file
inquire (unit=12, name= name)                     ! inquiry by unit
inquire (iolength= rl) y, z(1:7)           ! inquiry by output list
```

Inquiry by output list

The value assigned to the IOLENGTH= specifier is always processor-dependent. But this value may be reused as an input value for the RECL= specifier in an OPEN statement if a file with this record length is to be opened for direct access.

Inquiry by unit

The unit need not exist and need not be connected to a file. But if the unit is connected to a file, not only the properties of the unit are inquired about but also the properties of the file.

Inquiry by file

FILE= specifier: The file need not exist and need not be connected to a unit. The file name must be of a form acceptable to the processor. Trailing blanks are ignored.

NAME= specifier: The value assigned to the NAME= specifier may be different from the file name specified in the FILE= specifier. But its value may be reused as an input value for the FILE= specifier in an OPEN statement.

11.6.5 File Positioning Statements

There are three statements for the manipulation of external sequential files. The BACKSPACE statement may be used to position a file at most one record backwards. The REWIND statement may be used to position a file at its initial point. And the ENDFILE statement writes an endfile record.

BACKSPACE (UNIT = u [, IOSTAT = status_variable])

REWIND (UNIT = u [, IOSTAT = status_variable])

ENDFILE (UNIT = u [, IOSTAT = status_variable])

The **u** is a scalar nonnegative integer expression specifying the number of the unit connected to the file which is to be positioned or to which an endfile record is to be written. The **status_variable** must be scalar and of type default integer. It becomes defined with the value zero if no error condition occurs during the execution of the positioning statement; otherwise it becomes defined with a processor-dependent positive error status value.

The specifiers in the file positioning statements may be supplied in any order.

When an error condition occurs during the execution of a positioning statement that does not include an IOSTAT= specifier, program execution is terminated.

11

BACKSPACE statement

Execution of a BACKSPACE statement causes the file that is connected to the specified unit to be positioned before the current record if the file is currently positioned within a record. If there is no current record, the file will be positioned before the preceding record. If the preceding record is the endfile record, the file will be positioned before the endfile record.

If the most recent input/output statement for that file is a data transfer output statement and no positioning statement, an endfile record is implicitly (that is, automatically) written before the file is backspaced before the last record.

The BACKSPACE statement must not be executed for a nonexisting file. Backspacing over such records that are written using list-directed formatting is prohibited.

REWIND statement

Execution of a REWIND statement causes the file that is connected to the specified unit to be positioned at its initial point. If the file is currently positioned after the endfile record, it will also be rewound before the first record of the file.

The REWIND statement may be executed for a nonexisting file.

ENDFILE statement

Execution of an ENDFILE statement causes an endfile record to be written as the next record to the file. The file is positioned after this endfile record; that is, the endfile record becomes the last record of the file. Additional READ, WRITE, PRINT, or ENDFILE statements must not be executed for this file until a BACKSPACE or REWIND statement has been executed for the unit connected to this file.

The ENDFILE statement must not be executed for a file which is opened with ACTION="read".

```
real, dimension (12, 10) :: a, b
integer :: back_stat
⋮
do l=1,12
  write (unit=17, fmt=*) a(l, 1:10)
enddo
⋮
rewind (unit=17)                        ! <-- REWIND
do l=10,1,-1
  read (unit=17, fmt=*) b(l, 10:1:-1)
enddo
endfile (17)                            ! <-- ENDFILE
backspace (unit=17, iostat= back_stat)  ! <-- BACKSPACE
if (back_stat /= 0) then
   print *, "backspacing error"
endif
```

12 records are written to the file connected to unit 17. Then the file is rewound at its initial point. The first 10 records are read, and an endfile record is written after the 10th record. After the execution of the ENDFILE statement, the remaining two records containing the values $a(11, 1:10)$ and $a(12, 1:10)$ are no more available. The effect is as though they were deleted.

12   FORMATS

An explicit *format specification* (often abbreviated to "format") is used in conjunction with certain formatted input/output statements to control the input of external data and the output of internal data. During this data transfer, data are converted from the internal representation (in the storage) to the external character representation (in the record) or vice versa.

During the execution of a formatted input/output statement that includes such an explicit format specification, one *edit descriptor* is needed for each effective input/ output list item. Exception: two edit descriptors are needed for an input/output list item of type complex.

Edit descriptors are *data edit descriptors* and *control edit descriptors*.

The format specification appears as **format** within the data transfer statement.

12.1   Format Specification

A **format specification** is a list of format items enclosed in parentheses. These format items may be single edit descriptors or groups of edit descriptors. A group of edit descriptors is a list of format items enclosed in parentheses. A single repeatable edit descriptor or a group of edit descriptors may be preceded by a *repeat specification*. A comma is used to separate format items in a format specification.

12

A named constant must not appear within the character string forming the format specification. A format specification may be empty, such as ( ); but the parentheses must appear.

Leading blank characters preceding the left parenthesis of a format specification are allowed. No other characters must appear before or after a format specification. Unnecessary blank characters may appear within a format specification only on either side of a comma, either side of a parenthesis, after a repeat count, before a field width, or after a tabulator count.

If the **format** in a formatted input/output statement is a character array object, the format specification may be longer than the first array element. In this case, the format specification continues with the next array elements as though all array elements were chained in array element order.

```
character (len=22), save :: coform = "(tr15,f10.2,i5,es20.5)"
read (unit=9, fmt= coform) a, n, b
```

has the same effect as

```
character (len=6), dimension (4), save :: &
                 ff = (/ "(tr15,","f10.2,","i5,es2","0.5)  " /)
read (unit=9, fmt= ff) a, n, b
```

Variable format specification

If the character entity **format** is not a character constant expression, it is called a "variable format specification". In this case, the format specification need not be known until the execution of the program. For example, the program may compute and generate a suitable format specification.

A variable format specification must be complete when the formatted input/output statement is executed. The format specification for an internal WRITE statement must not be contained within the internal file. An input list item must not contain any part of the format specification. Note that a variable format specification may be redefined during the execution of the program.

```
(f7.2,es9.1,i5,f9.3,2es12.3,i6)    ←− Input record
12345678911234567892123456789311   ←− Position
```

This input record may be read into a character variable. And the character variable may be used as the FMT= specifier **format** in a data transfer statement:

```
character (len=31) :: frm
read (unit=12, fmt="(a)") frm
write (unit=15, fmt= frm) a, b, i, c, d, e, j
```

This WRITE statement has the same effect as

```
write(unit=15, fmt="(f7.2,es9.1,i5,f9.3,2es12.3,i6)") &
                                    a, b, i, c, d, e, j
```

12.2   Interaction between Input/Output List and Format

The interaction between an input/output list and a format specification is a dynamic process which is called the **format control**. Format control processes the input/output list and the format specification from the left to the right, associating in order one effective input/output list item with one corresponding data edit descriptor (exception: complex editing).

When format control encounters a control edit descriptor, the edit descriptor is executed immediately because an input/output list item is not needed.

The format specification must include at least one data edit descriptor if the input/output list includes an effective list item. An empty format specification of the form ( ) may occur only if the input/output list is empty or if all input/output list items are noneffective. If there is no current record, an advancing READ statement with empty format specification skips one record without transferring data; and an advancing output statement with empty format specification writes an empty record. If there is a current record, the execution of a nonadvancing data transfer statement with empty format specification does not change the file position within the current record.

Even if format control has already processed all input/output list items, format control continues processing remaining control edit descriptors (but no data edit descriptors). If there is no unprocessed effective input/output list item left, format control terminates when the next data edit descriptor, the right parenthesis at the end of the format specification, or a colon edit descriptor is encountered.

12.2.1 Repeat Specification, Groups of Edit Descriptors

A single repeatable edit descriptor or a group of edit descriptors (enclosed in parentheses) may be preceded by a **repeat specification**, which must be an unsigned nonzero integer literal constant without an explicitly specified kind type parameter. Such a repeat specification n has the effect as though n identical edit descriptors or n identical groups of edit descriptors were specified.

12

The format specification (2i4,2(i5,tr2,es7.2)) has the effect of the format specification (i4,i4,i5,tr2,es7.2,i5,tr2,es7.2). These different forms with or without a repeat specification are equivalent during the execution of the program as long as no *reversion of format control* occurs (see below).

A group of edit descriptors may include another group of edit descriptors (enclosed in parentheses). The maximum nesting level of such groups is processor-dependent.

12.2.2 Reversion of Format Control

If format control encounters the right parenthesis at the end of the format specification and there is another unprocessed effective input/output list item, **reversion of format control** occurs as follows: the current record is terminated, the file is positioned at the beginning of the next record, and format control reverts to the beginning of the last preceding group of edit descriptors. If this group is preceded by a repeat specification, this repeat specification will also be reused. If the format specification does not include such a group, format control reverts to the beginning of the format specification. In any case, format control continues to process edit descriptors from the left to the right beginning with the found position. And if format control encounters the last right parenthesis of the format specification for a second time, reversion of format control occurs again, and so on.

Reversion of format control has no effect on sign control (S, SP, and SS editing).

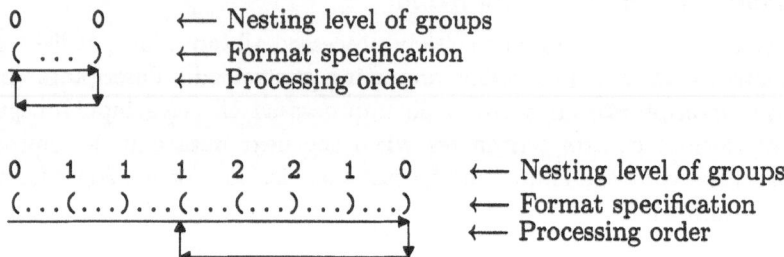

The marked group, where format control continues after reversion, is reused including its repeat specification (if any).

```
character (len=20), parameter :: format = "(i4,2(i3,f8.2),i5)"
read (unit=11, fmt=format) i, j, a, k, b, l, m, c, n, d
```

The data edit descriptors i4, i3, f8.2, i3, f8.2, and i5 are used in this order for the transmission of the input list items i, j, a, k, b, and l. Then the next record is read, and the data edit descriptors i3, f8.2, i3, and f8.2 are used to transmit the remaining input list items m, c, n, and d.

12.3   Edit Descriptors

There are two kinds of edit descriptors: *data edit descriptors* and *control edit descriptors*.

Data edit descriptors are used to describe the editing, that is, the kind of the conversion and the external representation of the data being transferred between input/output list items and a file.

Control edit descriptors are used to position in a record, to specify sign control for numeric output fields, or to terminate a record. Though no input/output list items are involved, control edit descriptors may affect subsequent data edit descriptors.

Data edit descriptors and the slash edit descriptor are **repeatable edit descriptors**. Control edit descriptors, except the slash edit descriptor, are **nonrepeatable edit descriptors**.

Data edit descriptors					
Descr.	Type	Descr.	Type	Descr.	Type
A	character	**ES.d**	numeric	**Lw**	logical
Aw	character	**ES.dEe**	numeric		
		Fw.d	numeric		
		Iw	numeric		
		Iw.m	numeric		

In the above table, upper-case letters specify the type of the editing. The lower-case letters **w**, **d**, **e**, and **m** are unsigned integer literal constants without an explicitly specified kind type parameter. They have the following meaning:

w specifies the *field width* of the external representation. The **field width** is the number of all characters transferred during the processing of the data edit descriptor including leading blank characters, sign, decimal point, and exponent, if any. **w** must be nonnegative.

d specifies the number of digits in the *fractional part* to the right of the decimal point. **d** may be zero.

e specifies the number of digits in the *exponent part*. **e** must be positive.

m specifies that at least **m** digits are written to the output field. **m** may be zero and must be $\leq$ **w**.

The valid values of **d**, **e**, and **m** depend on the value of the field width **w**.

12

Control edit descriptors					
Descr.	Type	Descr.	Type	Descr.	Type
SP	sign control	**Tn**	tabulator	**:**	format control
SS	sign control	**TRn**	tabulator	**/**	end of record
S	sign control	**TLn**	tabulator		

n is an unsigned positive integer literal constant without an explicitly specified kind type parameter.

Numeric editing

The data edit descriptors ES, F, and I may be used to specify the editing of numeric data.

Numeric input: Blank characters are ignored. A field containing only blank characters is interpreted as the value zero.

On input with an ES or F edit descriptor, a decimal point in the input field overrides the **d**, which otherwise would specify the position of the decimal point.

The input field may contain more digits than the F processor uses to approximate to the value.

Numeric output: Values of type real and complex may be rounded. The rounding method is processor-dependent.

The output representation is right-justified in the field. If the number of characters produced by the editing is less than the field width **w**, leading blank characters are transferred in the field. If the edit descriptor **Iw.m** is used to edit a "small" value, at least **m** digits are transferred in the field, possibly with leading zeros. Then, the field is filled with (**w** − **m**) leading blank characters.

A zero internal value or a positive internal value may be prefixed with a plus sign. Sign control depends on the F processor and on S, SP, or SS edit descriptors. A negative internal value is always prefixed with a minus. The zero internal value is never written with a leading minus.

Field width: On output, the number of transferred characters should not be greater than the field width **w**; and when the edit descriptor **ESw.dEe** is used, the number of digits produced for the exponent should not be greater than **e**. Otherwise, the field or the exponent part are too small, and the F processor fills the entire field with **w** asterisks.

Complex editing: A complex datum is represented (internally) as a pair of real data. Therefore, the editing of a complex datum is specified by two ES or F edit descriptors. The first of the two edit descriptors directs the editing of the real part and the second one directs the editing of the imaginary part. The two edit descriptors may be different. Additional control edit descriptors may appear between the two data edit descriptors.

```
complex :: a, b, c, d
read (unit=11, fmt="(2(es9.2),f7.3,es9.3,2(f5.1,f5.2))") a, b, c, d
```

Summary:

Edit descriptors		
Data	Control	
A ES F I L	/	S SP SS T TL TR :
repeatable	nonrepeatable	

12.3.1   A Edit Descriptors: A   Aw

The A edit descriptors are used to describe the editing of character data. The corresponding input/output list items must be of type character.

Input:

If the field width **w** is less than the length of the corresponding input list item, the input characters are transferred left-justified in the input list item, which is then filled with blank characters. If **w** equals the length *len* of the input list item, exactly *len* characters are transferred. If **w** is greater than the length *len* of the input list item, only the last *len* characters of the input field are transferred and the other (that is, the first $(\mathbf{w} - len)$) characters are ignored.

If **w** is not specified, the field width is given by the length *len* of the corresponding input list item; that is, *len* characters are transferred.

```
character (len=9) :: c
character (len=5) :: cc
read (unit=12, fmt="(a7)") c
read (unit=12, fmt="(tr8,a)") cc
```

```
circuit noises        ←— 1st input record
circuit breaker       ←— 2nd input record
123456789112345       ←— Position
```

After the execution of the READ statement, the variable c is defined with the value `circuit`␣␣ and the variable cc with the value **break**.

Output:

If the field width **w** is less than the length *len* of the corresponding output list item, only the first **w** characters are transferred. The other $(len - \mathbf{w})$ characters are ignored. If the field width **w** equals the length *len* of the output list item, exactly *len* characters are transferred. If **w** is greater than the length *len* of the output list item, the *len* characters are transferred right-justified in the output field, which is then filled with $(\mathbf{w} - len)$ leading blank characters.

If the field width **w** is not specified, the field width is given by the length of the corresponding output list item; that is, all characters of the output list item are transferred.

```
character (len=15), save :: cccc = "circuit breaker"
write (unit=18, fmt="(a7)") cccc
write (unit=18, fmt="(a18)") cccc
```

The WRITE statements create the following output records:

```
circuit                    ←— 1st output record
   circuit breaker         ←— 2nd output record
123456789112345678         ←— Position
```

12.3.2   Colon Edit Descriptor: :

The colon edit descriptor is used to terminate format control if there are no more effective input/output list items when the colon edit descriptor is encountered. The colon edit descriptor is ignored if there are more effective input/output list items.

12.3.3   ES Edit Descriptors: ESw.d   ESw.dEe

The ES edit descriptors are used to describe the editing of real and complex data. The corresponding input/output list items must be of type real or complex.

Input:

The input field consists of one or more subfields:

Integer part	Fractional part	Exponent part

Integer part: An optional sign followed by a sequence of digits. This sequence of digits may be omitted if the fractional part contains at least one digit.

Fractional part: An optional decimal point optionally followed by a sequence of digits. This sequence of digits may be omitted if the integer part contains at least one digit. If the decimal point is omitted, the **d** determines the number of digits in the fractional part. The integer part and the fractional part must contain together at least one digit.

The basic form of a real input datum consists of an integer part and a fractional part. This basic form may be optionally followed by an exponent part.

Exponent part: There are the following forms for the exponent part:

- A sign, followed by one or more optional blank characters, followed by a sequence of digits (possibly with intervening blank characters).

- The **exponent letter** e (or E, d, or D), followed by one or more optional blank characters, followed by an optional sign, followed by a sequence of digits (possibly with intervening blank characters).

If the input field contains a decimal point, the specification of **d** is ignored. If the decimal point is omitted, the **d** designates the position of the "imaginary" decimal point of the input value; that is, the input value is multiplied by 10^{-d}. On input with **ESw.dEe**, the specification of **e** is ignored.

Input field	Data edit descriptor	Value	Remarks
-534.9e-3	es9.1	−0.5349	All parts are complete, with sign; the exponent part begins with the letter e.
-35.24e2	es8.2	−3524.0	The sign of the exponent part is assumed to be +; the exponent part begins with the letter e.
-23.725+2	es9.3	−2372.5	The exponent letter is omitted, therefore, the sign of the exponent must be present.
19984.	es6.0	19984.0	The fractional part contains only the decimal point; the exponent part is omitted; the omitted sign of the integer part is assumed to be +.
.456739	es7.6	0.456739	The integer part and the exponent part are omitted; the omitted sign of the integer part is assumed to be +.
36988	es5.2	369.88	The decimal point is inserted between the 2nd and 3rd digit from the right; the omitted sign of the integer part is assumed to be +.

Output:

An ES edit descriptor creates an output field in scientific notation, such that the absolute value of the significand is ≥ 1 and < 10 (if the output value is nonzero).

Value	Format items	Output field
21.234	ss, es11.3	2.123e+01
80246.7	ss, es10.4	8.0247e+04
−0.72	ss, es12.2	-7.20e-01
0.000472	ss, es14.1e4	4.7e-0004
		12345678911234 ⟵ Position

The external character representation of a real value consists of a **significand** followed by an **exponent**. More precisely: the output field consists of leading blank characters (if necessary), followed by a sign (if necessary), followed by one digit, followed by the decimal point, followed by the remaining **d** most significant digits of the rounded value, followed by four or (**e** + 2) characters for the exponent.

The following details of the output field are processor-dependent: the plus sign for a positive value, and the form of the exponent if the absolute value of the exponent is ≤ 99.

12.3.4 F Edit Descriptor: Fw.d

The F edit descriptor is used to describe the editing of real and complex data. The corresponding input/output list item must be of type real or complex.

Input:

The form and the interpretation of the input field are the same as those of the ES edit descriptor.

```
read (unit=12, fmt="(f10.2,f15.6,f9.3)") a, b, c
```

Output:

The form of the output field is as follows: leading blank characters (if necessary), followed by a sign (if necessary), followed by a sequence of digits (if necessary), followed by the decimal point, followed by d digits.

Value	Format items	Output field	Remarks
45.783	ss, f6.3	45.783	
45.783	ss, f11.3	45.783	
-45.783	ss, f6.3	******	field width too small
0.45783	ss, f6.3	.458	possibly with leading zero
45.783	ss, f6.0	45.	

12345678901 ⟵ Position

The following details of the output field are processor-dependent: the plus sign for a positive value, and the leading zero before the decimal point if the absolute value is less than 1 (and at least a second digit is written).

12.3.5 I Edit Descriptors: Iw Iw.m

The I edit descriptors are used to describe the editing of integer data. The corresponding input/output list items must be of type integer.

Input:

The input field must contain (apart from blank characters) a character string having the form of an integer literal constant. The plus sign may be omitted. The specification of m is ignored; that is, on input, an Iw.m edit descriptor is treated as an Iw edit descriptor. w must be positive.

```
read (unit=18, fmt="(i3,i7,i5,i3,i2,i4)") i, j, k, l, m, n
```

539 -34   27 9   4 7   ←— Input record
1234567891123456789921234   ←— Position

The execution of the READ statement defines the variables i, j, k, l, m, and n
as follows: i = 539, j = -34, k = 27, l = 9, m = 0, and n = 47.

Output:

The form of the output field is as follows: leading blank characters (if necessary),
possibly a sign, followed by a sequence of digits which has the form of an unsigned
integer literal constant.

The form of the output field is processor-dependent, because the output of the
plus sign is partly processor-dependent.

On output with an **Iw.m** edit descriptor, if the value to be transferred occupies
fewer than **m** positions, the value is transferred right-justified with leading zeros
as a decimal value with **m** digits. And the leading (**w** − **m**) positions of the
output field are filled with blank characters.

On output of the value zero with an **Iw** edit descriptor, the digit zero is transferred right-justified with (**w** − 1) leading blank characters. On output of the value
zero with an **Iw.m** edit descriptor, **m** zeros are transferred right-justified; if **m** is
zero, no digits but **m** blank characters are transferred, giving a field of all blank
characters. When **m** and **w** are both zero on output of the value zero, one blank
character is transferred to the output field.

12

```
integer, save :: i = 57, j = 7694327, k = -3976
write (unit=16, fmt="(i7.4,i10,i8)") i, j, k
```

The WRITE statement creates the following record:

   0057   7694327   -3976   ←— Output record
1234567891123456789212345   ←— Position

12.3.6   L Edit Descriptor: Lw

The L edit descriptor is used to describe the editing of logical data. The corresponding input/output list item must be of type logical.

Input:

The form of the input field must be as follows: optional blank characters, optionally followed by a decimal point, followed by the letter "t" for a *true* value or
by the letter "f" for a *false* value, optionally followed by any other characters,
which are ignored.

```
logical :: lx, ly, lz
read (unit=14, fmt="(l5,l10,l7)") lx, ly, lz
```

```
    t    .false.    tru    ←— Input record
1234567891123456789212    ←— Position
```

The execution of the READ statement defines the variables lx, ly, and lz as follows: lx = .true., ly = .false., and lz = .true..

Output:

The form of the output field is as follows: $(w-1)$ blank characters, followed by the letter "t" or "f" depending on whether the value of the corresponding output list item is *true* or *false*.

```
logical, save :: la = .false., lb = .false., lc = .true.
write (unit=18, fmt="(l5,l8,l3)") la, lb, lc
```

The WRITE statement creates the following record:

```
    f        f   t    ←— Output record
1234567891123456    ←— Position
```

12.3.7   Sign Control Edit Descriptors: S SP SS

The S, SP, and SS edit descriptors are used to control the output of plus signs in numeric output fields. These edit descriptors may affect the "normal" processor-dependent way of producing a plus sign in numeric output fields.

SP: Positive values in subsequent numeric output fields are represented with a leading plus sign.

SS: Positive values in subsequent numeric output fields are represented without a leading plus sign.

S: The normal processor-dependent way of producing a plus sign for positive numeric output values is reestablished.

At the beginning of the execution of a formatted output statement, sign control is processor-dependent. The effect is as if format control has encountered an S edit descriptor as the first edit descriptor in the format specification.

```
integer, save :: i = 165
real, save    :: a = 1234.56, d = 67.4e-3
write (unit=16, fmt="(sp,i5,f10.2,ss,f10.2)") i, a, d
```

The WRITE statement creates the following record:

```
 +165   +1234.56        .07    ←— Output record
1234567891123456789212345    ←— Position
```

12.3.8   Slash Edit Descriptor: /

The slash edit descriptor indicates the end of the data transfer in or from the current record.

Sequential input: When format control encounters a slash edit descriptor, the remaining portion of the input record is skipped without transferring data, and the file is positioned at the beginning of the next record. This record becomes the current record. That is, if there are more input list items, data may be transferred in these list items from this subsequent record; otherwise this subsequent record is skipped without transferring data.

```
read (unit=14, fmt="(3i5//2i10)") i, j, k, l, m
```

```
  453    33 2876   112    ←── 1st input record
15432 2345    7 3418      ←── 2nd input record
   44     2  378     4    ←── 3rd input record
12345678911234567892      ←── Position
```

The execution of the READ statement defined the variables i, j, k, l, and m as follows: i = 453, j = 33, k = 2876, l = 44, and m = 2.

Direct input: When format control encounters a slash edit descriptor, the record number is increased by one and the file is positioned at the beginning of the next record. This record becomes the current record. That is, if there are more input list items, data may be transferred in these list items from this subsequent record; otherwise this subsequent record is skipped without transferring data.

Sequential output: When format control encounters a slash edit descriptor, the current record is terminated and a new record is created. This new record becomes the current record. It is the last record in the file (at this time). In the case of an external file, if the current record is empty when the slash edit descriptor is encountered, an empty record is created; in the case of an internal file, the record is filled with blank characters.

Direct output: When format control encounters a slash edit descriptor, the record number is increased by one and the file is positioned at the beginning of the record with this new record number. This record becomes the current record. That is, if there are more output list items, the output data may be transferred in this current record. If the current record is empty when the slash edit descriptor is encountered, it is filled with blank characters.

12.3.9   Tabulator Edit Descriptors: Tn TLn TRn

The T, TL, and TR edit descriptors are used to specify the position at which the next character will be transferred in or from the current record.

The T and TL edit descriptors may be used to specify a position backward from the current position. Positioning backward to a position which has been processed already is allowed until the *left tab limit* is reached. The left tab limit is given by the position of the current record immediately at the beginning of the execution of the formatted data transfer statement. Note that this may be a position which is *not* at the beginning of the current record. If an input/output statement transfers more than one record, the left tab limit is given by the first character position of each current record (possibly except the first one).

Tn: The transfer of the next character is to occur at the character position **n**. In this case, the character positions are counted beginning at the left tab limit.

TLn: The transfer of the next character is to occur at the character position **n** characters backward from the current position. If this position is backward from the left tab limit, the transfer of the next character is to occurs at the left tab limit.

TRn: The transfer of the next character is to occur at the character position **n** characters forward from the current position.

Input:

The characters in an input record may be transferred more than once (optionally with different edit descriptors). Even such a character position may be specified that is beyond the end of the record; in this case, no characters must be transferred from this position of the input record.

```
read (unit=18, fmt="(f10.6,t1,f10.3)") a, b
```

```
   45824569     ←— Input record
1234567890     ←— Position
```

The execution of the READ statement defines the variables a and b as follows: a = 45.824569 and b = 45824.569.

Output:

Suppose, format control has encountered a T, TL, or TR edit descriptor, when characters are transferred to or after the specified position, those character positions are filled with blank characters which are skipped without being previously filled.

```
write (unit=16, fmt="(a8,tr5,a10,t2,a4)") &
                             " railway", "racecourse", " run"
```

The WRITE statement creates the following record:

```
   runway      racecsource      ←— Output record
12345678911234567892123      ←— Position
```

13 PROGRAM UNITS AND SUBPROGRAMS

A **program unit** is a *main program* or a *module*. An F **program** consists of one main program and none or more modules. A program unit is a physically complete sequence of statements and comments which begins with a PROGRAM statement or MODULE statement and ends with an END PROGRAM statement and END MODULE statement, respectively.

Even if an F processor is able to compile program units separately, the F processor may not be able to compile them independently. For instance, the units in a program must appear in such an order that a module is already processed by the F processor when a USE statement with a reference to the module is processed.

During execution of a program, at first the main program takes over control by executing its first executable statement. The main program may invoke subprograms. A subprogram may invoke other subprograms and, under certain circumstances, a subprogram may invoke itself. If no error condition occurs, program execution is terminated when the END PROGRAM statement is executed or when a STOP statement (in the main program or in a module subprogram) is executed.

A module may be used to put related declarations, specifications, and definitions (also subprogram definitions) together. A module is the only program unit which is allowed to contain derived type definitions, generic interface blocks, and user-defined subprograms.

A **subprogram** is a *subroutine* or a *function*. Subprograms are program lines which belong together, which may be referenced in other parts of the program, and which solve particular parts of the complete problem solution of the program.

Subprograms written in the F language are defined as module subprograms. In addition to module subprograms, there are *external subprograms* and *intrinsic subprograms*.

External subprograms are written in another language than F, for instance in Fortran 90 or Fortran 95. Intrinsic subprograms are prepared by the F processor. Both external subprograms and intrinsic subprograms are ready for use. Intrinsic subprograms may be used in the entire program. An external subprogram may be used only where an appropriate interface block with an interface definition for the external subprogram is available.

Main programs, subroutines, and functions are executable. Modules are nonexecutable although a module may contain executable subprograms, the so-called *module subprograms*.

13.1 Main Program

A **main program** is a program unit whose first statement is a PROGRAM statement.

PROGRAM name	← PROGRAM statement
⋮	← Specification part; may be omitted
⋮	← Execution part; may be omitted
END PROGRAM name	← Executable END statement

The first statement in a main program must be a PROGRAM statement. The last statement must be an END PROGRAM statement. Additional PROGRAM statements must not appear. In addition to that, the scoping unit of a main program must not contain MODULE, FUNCTION, SUBROUTINE, or RETURN statements.

```
program sample
!----------------------------------------------------------------
  use sample_module                    !         specification part
  :
  real, dimension (4, 10) :: m1, m2, m3
!----------------------------------------------------------------
  open (unit=15, file=..., form=...)   !            execution part
  call form(m1, m2)
  :
  if (m1(3, 4) >= 78.95) then
    print *, m2(3, 4)
  endif
  :
end program sample
```

PROGRAM statement: The PROGRAM statement is used to specify a program name. This **name** is a global name which must not be used as a local name within the main program and even within the whole program.

Specification part: INTENT, OPTIONAL, PRIVATE, PUBLIC attributes, initial values, type definitions, and interface blocks must not be specified within a main program.

Execution part: The execution part contains the executable statements and executable constructs of the main program.

END PROGRAM statement: The **name** of the main program must be specified in the END PROGRAM statement which is the physically last line of the main program.

13.2 Modules

A **module** is a nonexecutable program unit which contains only declarations, specifications, and/or definitions (also subprogram definitions). The *public* entities in a module may be made *accessible* to other scoping units. An entity is **public** if it has the PUBLIC attribute.

A reference to a module is accomplished by a USE statement. The effect of such a module reference is that the public entities of the module become accessible to the scoping unit containing the USE statement. A module may reference other modules, but it must neither directly nor indirectly reference itself.

MODULE name	←— MODULE statement
⋮	←— Specification part
⋮	←— Subprogram part; may be omitted
END MODULE name	←— Nonexecutable END statement

The first statement in a module must be a MODULE statement. The last statement must be an END MODULE statement. Additional MODULE statements must not appear in a module. A module must not include a PROGRAM statement.

There are two sorts of modules. The first one *collects* definitions, specifications, and declarations from other modules into a package such that these module entities could be "used" collectively outside the module. The second one *creates* definitions, specifications, and declarations which are partly private, i.e. accessible only within the module, and partly public, i.e. accessible from outside the module.

A module of the first sort contains in its specification part only USE statements and a PUBLIC statement (without a list). The subprogram part is absent.

MODULE name	←— MODULE statement
USE ...	←— USE statement
⋮	←— More USE statements; may be omitted
PUBLIC	←— PUBLIC statement
END MODULE name	←— Nonexecutable END statement

A module of the second sort *must* be used if an F program defines derived types, generic names, operators, assignments, or subprograms written in F. In addition to that, such a module *may* contain declarations of variables and named constants. The form of such a "private" module is:

MODULE name   ← MODULE statement

⋮   ← Specification part

CONTAINS   ← May be omitted together with all subsequent
                           module subprograms

⋮   ← Module subprograms

END MODULE name   ← Nonexecutable END statement

MODULE statement: The **name** of the module is used in the program where public entities from the module are to be made accessible (by an appropriate USE statement). The name is a global name. It must not be used as a local name within the module. The name of the module must also be specified in the END MODULE statement.

```
module list
  real, public, dimension (10) :: a, e
  character (len=17), private  :: card
end module list
```

Specification part: If the specification part of a module contains any USE statement, it must also contain a PRIVATE or PUBLIC statement (without a list) which specifies the accessibility of all those module entities being accessed from other modules. A module without a USE statement must contain neither a PRIVATE nor a PUBLIC statement (without a list).

Within a module of the 1*st* sort, the statements must appear in the following order:

1. USE statement(s)

2. PUBLIC statement (without a list)

The PUBLIC statement has the effect that the entities being accessed from other modules are visible and can be "used" outside the module.

Within a private module, i.e. a module of the 2*nd* sort, the statements must appear in the following order:

1. USE statement(s): if necessary.

2. IMPLICIT NONE: may be omitted.

3. PRIVATE statement (without a list): is required if the module contains USE statements; otherwise the PRIVATE statement must be absent.

4. PRIVATE and PUBLIC statements (with a list): if necessary.

5. INTRINSIC statement(s): if necessary.

6. Type definitions, type declarations, and interface blocks: if necessary.

The PRIVATE statement (without a list) has the effect that the entities being accessed from other modules are available within this module only and cannot be accessed in this module from outside the module.

INTENT and OPTIONAL attributes must not appear in the specification part of a module. The names of the module subprograms may appear within the specification part of a module only in PRIVATE or PUBLIC statements. Executable statements must not appear in the specification part of a module.

For each module entity being specified, defined, declared in the module, or being accessed from other modules, one of the accessibility attributes PRIVATE or PUBLIC must be specified within the module:

- For each module entity accessed by USE association: by a PRIVATE or PUBLIC statement (without a list) depending on the sort of the module; otherwise,

- For declared data objects: within the type declaration statement;

- For defined derived types: within the TYPE definition statement;

- For defined module subprograms: by a PRIVATE or PUBLIC statement (with a list);

- For defined operators or extended intrinsic operators: by a PRIVATE or PUBLIC statement (with a list);

- For defined assignments: by a PRIVATE or PUBLIC statement (with a list); and

- For generic names: by a PRIVATE or PUBLIC statement (with a list). **13**

If a derived type is public but its type components are private, a PRIVATE statement (without a list) must appear within the type definition.

CONTAINS statement: The CONTAINS statement must be omitted if the module does not contain any module subprogram.

Module subprograms : A module of the 2*nd* sort may contain subprograms. These *module subprograms* appear subsequent to the specification part of the module between the CONTAINS statement and the END MODULE statement.

END MODULE statement: The **name** of the module must be specified in the END MODULE statement which is the physically last line of the module.

13.2.1 USE Statement

The USE statement is used to reference a module which has already been processed by the F processor. Where an appropriate USE statement appears, declarations, specifications, and definitions may become accessible from the referenced module; and we say that the scoping unit containing the USE statement has *access to the module*. The following entities can be made accessible from a module only if they are public: named variables, named constants, module subprograms, derived types, interface blocks, and generic identifiers defined by generic interface blocks (for instance, generic names, generic operators, and generic assignment).

USE module

USE module, local_name => use-name [, local_name => use-name]...

USE module, ONLY : only [, only]...

module is the name of a module, and each **local_name** is a local name in the scoping unit containing the USE statement. Each **use-name** is the name of a public entity in the **module**. And each **only** in the *only-list* is one of the following:

use-name
local_name => use-name
generic_name
OPERATOR (operator)
ASSIGNMENT (=)

The last three specifications designate corresponding generic interface blocks.

Entities in the scoping unit with the USE statement and public entities in the referenced module may become associated. This is called *USE association*.

A scoping unit may include more than one USE statement. These must be USE statements referencing different modules, if necessary. No two USE statements in a scoping unit are allowed to reference the same module, neither directly nor indirectly by a module referencing a previously referenced module.

If a USE statement *without* the keyword ONLY is processed, *all* public entities within the specified module are made accessible. A USE statement *with* the keyword ONLY makes *only* the specified public entities in the module accessible. In both cases, public named entities may be renamed for use in the scoping unit containing the USE statement by specifying a local name for them. The name of an intrinsic subprogram cannot be renamed.

Module entities *must* be renamed if there are name conflicts between entities accessed from different modules or between entities accessed from the module and entities in the scoping unit containing the USE statement.

`use graphic_lib`

All public entities in module `graphic_lib` are made accessible such that each named entity has the same name both in the scoping unit containing the USE statement and in the module.

```
use graphic_lib, cir => circle, ar => arc, li => line
```

As above, except the module entities **circle**, **arc**, and **line** are accessible by the local names **cir**, **ar**, and **li**, respectively, in the scoping unit containing the USE statement.

A local name of an entity made accessible by a USE statement must not be explicitly respecified in a type declaration statement or in another specification statement in the scoping unit containing the USE statement.

USE with ONLY

If a scoping unit needs only a part of a module, the keyword ONLY followed by an only-list may be specified in the USE statement. In this case, only those (public) entities are made accessible which are specified in the only-list. A USE statement which includes the keyword ONLY does not override a USE statement *without* the keyword ONLY.

```
use statistic_lib, only : gauss, hg => histogram
```

Only the module entities **gauss** and **histogram** are made accessible. These entities are identified by the local names **gauss** and **hg** in the scoping unit containing this USE statement.

13.2.2   Typical applications

Derived types: Nonintrinsic datatypes may be defined only within the specification part of a module. If such a derived type is needed outside the module, it must be defined with the PUBLIC attribute.

Data structures: Data structures (structures, structure objects) are scalar derived type objects. The type definition of this derived type must appear in the specification of a module. If the derived type has the PUBLIC attribute, it can be made accessible from outside the module where data structures of this derived type can be declared.

Generic interface blocks: Operator interface blocks, assignment interface blocks, and interface blocks with generic names may be specified only within the specification part of a module. Outside the module, defined operators, defined assignments, and generic names can be made accessible, if they are specified with the PUBLIC attribute.

Abstract datatypes: A derived type and its operators may be defined in a module and specified to have the PUBLIC attribute. For the implementations of these operators, i. e. for the associated module functions, the PRIVATE attribute may be specified. Such the derived type and its operators are visible and the implementations of the operators are nonaccessible outside the module.

Global data: In the specification part of a module, data objects may be declared. These objects have global character if they are public. Such an object exists physically only once even if it is "used" in several other scoping units.

Program library: The specification part of a module may contain interface blocks with interface definitions for external subprograms being collected in a program library. Where these interface blocks are available by USE association, the external subprogram may be referenced and the F processor can check whether the subprogram reference agrees with the interface definition.

F subprograms: A subprogram being written in F must be embedded as a module subprogram within the subprogram part of a module. Other F subprograms are not supported.

13.3   Subprograms

A **subprogram** is either a *function* or a *subroutine*. Functions and subroutines are different from each other with regard to subprogram definition and subprogram reference.

In addition to *user-defined* subprograms which are supplied by the programmer, there are *predefined* subprograms. The most important predefined subprograms are the *intrinsic subprograms*.

A subprogram is not an independent part of a program. It will be executed only when it is (explicitly or implicitly) referenced. The referencing scoping unit and the referenced subprogram may exchange information during the execution of the subprogram.

Classification of subprograms

Subprograms may be classified by their characteristics as follows:

- A **module subprogram** is a part of a module. If it is public, its subprogram definition may be made accessible to other program units or subprograms. A user-defined subprogram which is written in F is a module subprogram.

- An **intrinsic subprogram** is a predefined subprogram. It is provided by the F processor.

- An **external subprogram** is a user-defined subprogram which is not written in F but in another programming language, for instance, in Fortran 90 or Fortran 95.

An **operator function** is a module function which defines the interpretation of a defined operator or which defines the interpretation of an extension of an intrinsic operator.

An **assignment subroutine** is a module subroutine which defines the interpretation of a defined assignment.

A **dummy subprogram** is a dummy argument which is specified by an interface block as a subprogram and which may appear in a subprogram reference as the name of the referenced subprogram.

A function may be referenced only in an expression. The function reference may appear explicitly as an operand in the expression. But a function may also be implicitly (that is, automatically) invoked when a defined operator or an extended intrinsic operator is encountered during expression evaluation. A subroutine is either explicitly referenced and invoked by the execution of a CALL statement or it is implicitly invoked when a defined assignment statement is executed.

The name of a subprogram identifies the **subprogram interface**. A subprogram interface has the following characteristics:

- Sort of subprogram (function or subroutine);
- Specific name of the subprogram;
- Names of the dummy arguments;
- Sort of dummy arguments (datum or subprogram);
- Characteristics of the dummy arguments;
- Generic name of the subprogram (if any); and possibly
- Characteristics of the function result.

13

The execution of a subprogram written in F begins with the execution of the first executable statement subsequent to the FUNCTION or SUBROUTINE statement.

13.3.1 Module Functions

A function written in F begins (apart from optional leading comment lines) with a FUNCTION statement, ends with an END FUNCTION statement, contains between the FUNCTION and the END FUNCTION at least a type declaration statement for the result variable and one executable statement, and it may contain comment lines. It must not contain a MODULE, SUBROUTINE, or PROGRAM statement. Such a function must be embedded in a module; i.e., it is a module function. A module function is part of the module and has access to certain module entities by host association.

A module function may be referenced from any other module subprogram defined within this module. If the function has the PUBLIC attribute, it may become available outside the module by USE association and may be referenced there.

A function is referenced and invoked during expression evaluation either explicitly when the function name is encountered as an operand or implicitly when a defined operator or an extended intrinsic operator is encountered. The invoked function returns a value to the place of the function reference, this value is the function result. Apart from this function result, the argument list may be used to exchange information between the function and the invoking scoping unit. When a RETURN statement or the END FUNCTION statement is executed, the execution of the function is terminated, and control returns to the expression containing the function reference.

```
module move_lib
  public :: rotate, error
  ⋮
contains                              ! module subprograms
  real function rotate (angle) result (rot)
    real, intent(in) :: angle
    real             :: rot
    real             :: pi
    ⋮
    rot = cos(angle + pi)
    call error(x=rotate)              ! module subroutine reference
  end function rotate

  subroutine error (x)
    real, intent(in) :: x
    if (x < 0.0) then
      stop "X is negative"
    endif
  end subroutine error
end module move_lib

program plane
  use move_lib                        ! module reference
  ⋮
  xpoint = r * rotate(angle=phi)      ! module function reference
  ⋮
  call error(x=xpoint-10)             ! module subroutine reference
  ⋮
end program plane
```

13.3.1.1 Function Definition

[...] FUNCTION name ([...] **) RESULT (...)**   ← FUNCTION statement

:   ← Specification part

:   ← Execution part

END FUNCTION name   ← Executable END statement

Within a function definition, the specification statements in the specification part must appear in the following order:

1. USE statement(s): if necessary.

2. INTRINSIC statement(s): if necessary.

3. Interface blocks and type declarations with attributes for the dummy arguments: if any.

4. Type declaration with attributes for the result variable.

5. Other type declarations: if necessary.

A local variable declared in the specification part of a function must not have the SAVE attribute. This is why such a variable cannot be initialized.

The subprogram definitions of all module subprograms of a particular module appear between the CONTAINS statement and the END MODULE statement.

FUNCTION statement

13

The FUNCTION statement is used to specify the name of the function, the dummy argument list, and whether or not the function is recursive. The FUNCTION statement must be the first statement of a function.

[RECURSIVE] FUNCTION name ( ) RESULT (result)

[RECURSIVE] FUNCTION name (dummy_argument [, dummy_argument] ...)
                                                      RESULT (result)

The keyword RECURSIVE must be specified for a recursive function. **name** is the **specific name** of the function. An additional *generic name* may be specified by a generic interface block. Each **dummy_argument** is the name of a variable or the name of a dummy subprogram. Any optional dummy argument must follow all nonoptional dummy arguments. And **result** is the name of the *result variable*.

```
function mean (x) result (mw)
  real, dimension (:), intent(in) :: x
  real                            :: mw
  integer                         :: i, n
  n = size(x)
  mw = 0.0
  do i=1,100
    mw = mw + i/real(n)*x(i)
  enddo
  mw = mw/n
end function mean
```

Function result

A function, i. e. its result variable, has a type, (if applicable) a kind type para-meter, and (if applicable) a character length. These attributes must be specified explicitly in the specification part of the function. If a function is of a derived type, this type must be available within the specification part of the function by USE association or host association.

The result variable of a function may have additional attributes, such as the DIMENSION and the POINTER attribute which must be specified within the specification part of the function. The result variable of an array-valued non-pointer function must be an explicit-shape array. The array bounds may depend on nonconstant specification expressions.

On return to the invoking expression, the function result is given by the value of the result variable. This **result variable** is a local variable in the function. Any occurrence of the name of the function in the execution part of the func-tion is interpreted as a recursive reference to the function. The name of the function must not appear in type declaration statements or other specification statements within the scoping unit of the function. The result variable of a character function must not be specified to have character lenght *.

Nonpointer function: If the function result is *not* a pointer, the result variable must become defined with a valid value during the execution of the function. This value may be used or may be redefined during the execution of the function. Finally, when a RETURN statement or the END FUNCTION statement is executed, the current value of this result variable is the function result, which is then reused as an operand in the invoking expression.

Pointer function: If the function result is a pointer, the pointer result va-riable must become associated with a target during the execution of the function, or the association status of this pointer must become disassociated. The result variable of the function may be an array, that is, an array pointer. The shape of such a function result is determined by the shape of the result variable when control returns to the invoking expression.

Dummy arguments

Dummy argument names are local names. If they identify data objects, these data objects are specified more precisely by a type declaration statement within the specification part of the function.

Dummy arguments being arrays must be specified as arrays in the specification part of the function. If such a dummy argument is a nonpointer array, it must be specified as an assumed-shape array. Dummy arguments of type character must be specified with character length *.

Each dummy argument which identifies neither a pointer nor a subprogram must be specified with an INTENT(IN) attribute.

If a dummy argument has the OPTIONAL attribute, all subsequent dummy arguments in the dummy argument list must also have the OPTIONAL attribute.

Execution part

Any subprogram explicitly referenced from a function must also be a function. Implicit function references via defined operations and implicit subroutine references via defined assignments are allowed from the execution part of a function.

Any variable which is accessed by USE or host association, or is a dummy argument must not be used in the following contexts:

- As the left-hand side of an assignment statement;
- As an input list item in a READ statement;
- As an internal file in a WRITE statement;
- As an IOSTAT= specifier in an input/output statement;
- As the right-hand side of a pointer assignment statement;
- As the right-hand side of an assignment statement in which the left hand side has an ultimate pointer component;
- As an object or status variable in an ALLOCATE or DEALLOCATE statement;
- As an object in a NULLIFY statement; or
- As an actual argument for a corresponding dummy argument with the POINTER attribute.

Any subroutine referenced by a defined assignment and any subprogram invoked during such reference must obey the rules relating to variables in a function except that the first argument of the subroutine may have INTENT(OUT) or INTENT(INOUT).

A function must not contain one of the following statements:

- OPEN and CLOSE;

- BACKSPACE, REWIND, and ENDFILE;

- INQUIRE;

- READ and WRITE, except it refers to an internal file, or has UNIT=* and FMT= specifier, or is a READ statement with no control information list.

END FUNCTION statement

The **name** of the function must be specified in the END FUNCTION statement which is the physically last line of the function.

Recursive functions

The keyword RECURSIVE must be specified in the FUNCTION statement if the function directly or indirectly invokes itself.

```
recursive function fff (duar) result (f)
  real, dimension (:), intent(in) :: duar
  real, dimension (size(duar))    :: f
  integer :: n
  n  = size(duar)
  if (n <= 1) then
    f = duar
  else
    n = n/2
    f(:n)   = fff(duar(:n))
    f(n+1:) = fff(duar(n+1:)) + f(n)
  endif
end function fff
```

13.3.1.2 Explicit Function Reference, Invocation

A function is explicitly invoked by a function reference during expression evaluation. A function invocation returns a value for use in the expression at the place of the function reference; this value is the function result.

name ( )

name (aa [, aa]...)

The **name** is the (specific or generic) name of the function. And each **aa** is an actual argument specification, which may be written as a *positional argument* or as a *keyword argument*.

For a positional argument, only the actual argument is specified. A keyword argument has the form **dummy_argument = actual_argument**, where **dummy_argument** is the name of a dummy argument from the interface of the referenced function. This name is used as a keyword for the **actual_argument** which will become associated with this dummy argument.

A function defined with an empty dummy argument list must be referenced with an empty actual argument list; that is, the parentheses must be written. If the actual argument list includes a keyword argument, all subsequent actual arguments must also be specified as keyword arguments.

Characteristics of the **name** used in the function reference

In the scoping unit containing the function reference, additional specifications for the function result need not and must not appear because:

- If the function reference appears within the module containing the module function, the characteristics of the function are already known to all subprograms defined in the module, and
- If the function reference appears outside the module, the characteristics of the module function are made available by USE association in the scoping unit with the function reference.

```
program abc
  :
real, dimension (100) :: jan, feb
real                  :: minmw, ref
  :
minmw = mw(jan) * 100.0
ref   = mw(feb) * 10.0
if (ref < minmw) then
  minmw = ref
endif
  :
end program abc
```

Regardless of whether the referenced function **mw** is a module function or an external function, the main program must not contain a type declaration for the real entity **mw**.

Execution

The invocation of a function is executed as follows:

1. Evaluation of the actual arguments which are expressions with operators or which are enclosed in parentheses; and determination of the position of the actual arguments which are subobjects.

2. Association of the arguments.

3. Execution of the function beginning with the first executable statement in the function.

4. Return to the expression containing the function reference, where the function result is now available as an operand.

Recursive reference: For each invocation of a function, a separate **instance** of the function is created and executed. Each instance of a function has an independent set of dummy arguments and nonsaved local variables. Entities which are available by USE association or host association are shared and may be reused by all instances of the function. But there are certain restrictions with regard to the use of such entities within the function.

13.3.1.3 Operator Functions

An *operator function* is a special module function which defines the interpretation of a defined operator or of an extended intrinsic operator. It is automatically invoked when the operator is encountered during expression evaluation.

A function may be used as an operator function only if it has one or two non-optional dummy arguments, if the dummy arguments are data objects, and if the arguments are specified with INTENT(IN) attribute.

Whether or not such a function eventually is implicitly invoked as an operator function depends on additional conditions.

13.3.2 Module Subroutines

A subroutine begins (apart from optional leading comment lines) with a SUBROUTINE statement, it ends with an END SUBROUTINE statement, and may contain between the SUBROUTINE and the END SUBROUTINE statement additional executable statements, nonexecutable statements, and comments. A subroutine must not contain a MODULE, FUNCTION, or PROGRAM statement. Such a user-defined subroutine must be embedded in a module, i. e. it is a module subroutine. A module subroutine is part of the module and has access to certain module entities by host association.

A module subroutine may be referenced from any other module subprogram defined within this module. If the subroutine has the PUBLIC attribute, it may become available outside the module by USE association and may be referenced there (see the example at the beginning of 13.3.1).

A subroutine is referenced and invoked when a CALL statement is executed or when a defined assignment statement is executed. The argument list may be used to exchange information between the subroutine and the invoking scoping unit. When a RETURN statement or the END SUBROUTINE statement is executed, the execution of the subroutine is terminated, and control returns to the invoking scoping unit.

13.3.2.1   Subroutine Definition

[**RECURSIVE**] **SUBROUTINE name (...)**   ⟵ SUBROUTINE statement

    ⋮   ⟵ Specification part; may be omitted

    ⋮   ⟵ Execution part; may be omitted

END SUBROUTINE name   ⟵ Executable END statement

Within a subroutine definition, the specification statements in the specification part must appear in the following order:

1. USE statement(s): if necessary.

2. INTRINSIC statements(s): if necessary.

3. Interface blocks and type declarations with attributes for the dummy arguments: if any.

4. Other type declarations: if necessary.

The subprogram definitions of all module subprograms of a particular module appear between the CONTAINS statement and the END MODULE statement.

A subroutine which is referenced within a function must obey restrictions which are presented in 13.3.1.1.

SUBROUTINE statement

The SUBROUTINE statement is used to specify the name of the subroutine and the dummy argument list. The SUBROUTINE statement must be the first statement of a subroutine.

[**RECURSIVE**] **SUBROUTINE name ( )**

[**RECURSIVE**] **SUBROUTINE name (dummy_argument [, dummy_argument]...)**

The keyword RECURSIVE must be specified for a recursive subroutine. **name** is the **specific name** of the subroutine. An additional *generic name* may be specified by a generic interface block. Each **dummy_argument** is the name of a variable or the name of a dummy subprogram. Any optional dummy argument must follow all nonoptional dummy arguments.

```
subroutine summa (x, y, r)
  real, dimension (:), intent(in)  :: x, y
  real, dimension (:), intent(out) :: r
  r = x + y
  where (r > 775.0)
    r = 775.0
  end where
end subroutine summa
```

Dummy arguments

Dummy argument names are local names. If they identify data objects, these data objects are specified more precisely by a type declaration statement within the specification part of the function.

Dummy arguments being arrays must be specified as arrays in the specification part of the subroutine. If such a dummy argument is a nonpointer array, it must be specified as an assumed-shape array. Dummy arguments of type character must be specified with character length *.

Each dummy argument which identifies neither a pointer nor a subprogram must be specified with an INTENT attribute.

If a dummy argument has the OPTIONAL attribute, all subsequent dummy arguments in the dummy argument list must also have the OPTIONAL attribute.

END SUBROUTINE statement

The **name** of the subroutine may be specified in the END SUBROUTINE statement which is the physically last line of the subroutine.

Recursive subroutines

The keyword RECURSIVE must be specified in the SUBROUTINE statement if the subroutine directly or indirectly invokes itself.

13.3.2.2   Explicit Subroutine Reference, CALL Statement

A subroutine is explicitly invoked by the execution of a CALL statement.

CALL name ( )

CALL name (aa [, aa]...)

The **name** is the (specific or generic) name of a subroutine. And each **aa** is an actual argument specification, which may be written as a *positional argument* or as a *keyword argument*.

For a positional argument, only the actual argument is specified. A keyword argument has the form **dummy_argument = actual_argument**, where **dummy_argument** is the name of a dummy argument from the interface of the referenced subroutine. This name is used as a keyword for the **actual_argument** which will become associated with this dummy argument.

A subroutine which is defined with an empty argument list must be referenced by a CALL statement with an empty actual argument list. If the actual argument list includes a keyword argument, all subsequent actual arguments must also be specified as keyword arguments.

Execution

A CALL statement is executed as follows:

1. Evaluation of the actual arguments which are expressions with operators or which are enclosed in parentheses; and determination of the position of the actual arguments which are subobjects.

13

2. Association of the arguments.

3. Execution of the subroutine beginning with the first executable statement in the subroutine.

4. Return to the invoking scoping unit containing the CALL statement.

```
program su
⋮
real, dimension (100) :: jan, feb, y1993
read (unit=14) jan, feb
call terms(jan, feb, y1993)
write (unit=16) y1993
⋮
end program su
```

After return to the invoking scoping unit, the execution of the CALL statement terminates, and program execution is continued with the WRITE statement.

Recursive reference: For each invocation of a subroutine, a separate **instance** of the subroutine is created and executed. Each instance of a subroutine has an independent set of dummy arguments and nonsaved local variables. All other entities are shared by all instances of the subroutine. For example, if a saved variable is defined (with a valid value) in one instance of the subroutine, the variable may be reused with this value in any other instance of the subroutine.

13.3.2.3   Assignment Subroutines

An *assignment subroutine* is a special module subroutine which defines the interpretation of a defined assignment statement. It is automatically invoked when the defined assignment statement is executed.

A subroutine may be used as an assignment subroutine only if it has two nonoptional dummy arguments, if the dummy arguments are data objects, if the first dummy argument is specified with the INTENT(OUT) or INTENT(INOUT) attribute, and if the second dummy argument is specified with the INTENT(IN) attribute.

Whether or not such a subroutine eventually is implicitly invoked as an assignment subroutine depends on additional conditions.

13.3.3   External Subprograms

An external subprogram is a user-defined subprogram which is not embedded in a module, which is not written in F but in any other programming language, for instance, Fortran 90 or Fortran 95, and which is prepared ready for use.

Reference to an external subprogram

A reference to an external function or to an external subroutine has the same form and the same properties as a reference to a module function and module subroutine, respectively.

In the scoping unit with the reference to an external subprogram, a *specific interface block* with an interface definition of the external subprogram must be available. This interface block must be supplied in the specification part of a module and must be made accessible by USE association if the subprogram reference appears outside the module.

13.3.4   Dummy Subprograms

A dummy argument identifying a subprogram is called a **dummy subprogram**. In the specification part of the subprogram with such a dummy argument, a *specific interface block* with an interface defintion of the dummy subprogram must appear. Regardless of whether the dummy subprogram is explicitly referenced, it is determinable from the interface definition whether the dummy subprogram is a *dummy function* or a *dummy subroutine*.

Reference to a dummy subprogram

A reference to a dummy function or dummy subroutine has the same form and the same properties as a reference to a module function and module subroutine, respectively.

13.3.5   Interface Blocks

An **interface block** is used to specify the characteristics of one or more subprogram interfaces.

An interface block may contain the following specifications:

- One or more *interface definitions*, in which *all* characteristics of (one or several) external subprograms or (one or several) dummy subprograms are specified. This form of an interface block is called a *specific interface block.*

- A *generic name* and the names of module subprograms. This interface block is used to specify a (common) *generic name* for the subprograms whose *specific names* are specified in this interface block.

  This form of an interface block is a special form of a *generic interface block* and is called an *interface block with a generic name.*

- The designation of an operator and the names of those module functions which are specified to be the operator functions for this defined operator or extended intrinsic operator. This form of an interface block is a special form of a *generic interface block* and is called an *operator interface block.*

- The assignment operator and the names of those module subroutines which are specified to be the assignment subroutines for a defined assignment. This form of an interface block is a special form of a *generic interface block* and is called an *assignment interface block.*

13

An interface block may appear in the specification part of a subprogram and in the specification part of a module. It is a part of the specification part of its host scoping unit. The specifications of an interface block are accessible in the host scoping unit, but the specifications in the host scoping unit are *not* accessible in the interface block.

```
module graph
  public :: circle, link, rotate, segment
  interface                       ! interface block for
    subroutine circle (x, y, r)   ! external subprogram
      :
    end subroutine circle
  end interface
  interface link                  ! generic name
    module procedure segment      ! specific name
  end interface
contains                          ! module subprograms
  function rotate (angle) result (box)
    interface                     ! interface bl. in module function
      subroutine angle (c)        ! dummy subprogram
        :
      end subroutine angle
    end interface
    :                             ⟵ other specs in module function
    :                             ⟵ execution part of module function
  end function rotate
  subroutine segment (a, b)       ! module subroutine
    :
  end subroutine segment
end module graph

program mp
  use graph
  :                               ⟵ remaining spec. part of main prog.
  :                               ⟵ execution part of main program
end program mp
```

A **specific interface block** for external subprograms or for dummy subprograms has the following form:

INTERFACE

    :   ⟵ *Interface definition(s)*

END INTERFACE

Such a specific interface block enables the specified external or dummy subprograms to be referenced. An interface block appearing in the specification part of a subprogram must not contain an interface definition for this subprogram.

Interface definition: The interface block must contain at least one interface definition which begins with a FUNCTION or SUBROUTINE statement, which ends with an END FUNCTION and END SUBROUTINE statement, respectively, and which contains between these two statements additional specification statements that characterize the interface, that is, the function result (if any) and the dummy arguments.

An interface definition must contain attribute specifications only for data objects that are function results or dummy arguments. The interface definition must specify INTENT attributes for all dummy arguments except pointers and dummy subprograms.

If the interface block contains an interface definition of an external subprogram, the interface definition within the interface block must be consistent with the subprogram definition.

If the interface block contains an interface definition of a dummy subprogram, the characteristics of any subprogram being associated as an actual argument must agree with the interface definition within the interface block. There are two exceptions: the name of the dummy subprogram may be different from the name of the associated actual subprogram, and the names of the dummy arguments within the interface definition may be different from those in any associated actual subprogram.

An interface block with an interface definition for a dummy subprogram specifies **13** that the dummy argument is a function or a subroutine which has exactly the properties specified in the interface block.

The name of a dummy subprogram must not appear outside the interface block in any other specification statement within the specification part containing the interface block.

An **interface block with a generic name** has the following form:

INTERFACE generic_name
   MODULE PROCEDURE ...   ←— MODULE PROCEDURE statement(s)
   ⋮

END INTERFACE

The MODULE PROCEDURE statement is used to specify the names of certain module subprograms. It has the following form:

MODULE PROCEDURE name [, name]...

A generic interface block must specify either subroutines or functions but not both. Such an interface block specifies a **generic_name** for all those module subprograms which are specified in the interface block. These subprograms may then be referenced optionally by their specific names or by this generic name. If the generic name of a subprogram is available but the specific name is private, the subprogram may be referenced by its generic name, but it cannot be referenced by its specific name outside the module.

If an interface block with a generic name specifies more than one specific subprogram name, the generic name becomes "overloaded" by these specific names. The generic name must not be the same as any one of the specific names in the interface block. Note that two or more generic interface blocks accessible in a scoping unit are interpreted as a single generic interface if they have the same generic name.

Overloading of a generic subprogram name by specific subprogram names is allowed if each of the affected interfaces unambiguously matches exactly one subprogram at the time of the reference to this generic name. Overloading with an intrinsic name must not result in an ambigous subprogram reference (see below).

Operator interface blocks, assignment interface blocks

Operator interface blocks and assignment interface blocks are presented in context with defined operators and defined assignment statements in 7.5.1 and 8.2, respectively.

13.3.6   Overloaded Generic Subprogram Names

Normally, a user-defined subprogram is identified in a scoping unit by exactly one name. But in addition to this *specific name*, a module subprogram may have a *generic name*. Such a generic name of a module subprogram is specified in the INTERFACE statement of a generic interface block.

If an interface block with a generic name specifies two or more subprograms, the generic name is **overloaded** because it identifies more than one subprogram. Overloading of a generic subprogram name by specific subprogram names is allowed if unambiguously exactly one of the specified subprograms is affected when the generic subprogram name is used in a subprogram reference.

Therefore, any two such functions or two such subroutines having the same generic name in a scoping unit must be distinguishable from each other when they are referenced by their common generic name. They are distinguishable if one of them has more nonoptional dummy arguments of a particular data type, kind type parameter, and rank than the other has dummy arguments (including optional dummy arguments) of that data type, kind type parameter, and rank. They also are distinguishable if one of them has a nonoptional dummy argument that corresponds by position in the argument list to a dummy argument with different properties in the other, and if it has a nonoptional dummy argument that corresponds by name in the argument list to a dummy argument with different properties in the other. These properties of the corresponding dummy argument in the other argument list are: presence, data type, kind type parameter, or rank. The dummy argument that disambiguates by position must either be the same as or occur earlier in the argument list than the one that disambiguates by keyword.

Intrinsic subprograms

If the generic name specified in the INTERFACE statement of a generic interface block is the name of an intrinsic subprogram, the module subprograms specified in this generic interface block extend the predefined meaning of this intrinsic subprogram. In analogy to the above rule (concerning the unambiguity of the references to overloaded user-defined generic subprogram names), the references to all subprograms having this generic name must be unambiguous, as though the intrinsic subprogram would consist of a collection of "predefined module subprograms" and as though the names of these subprograms were also specified in the generic interface block in addition to the user-defined module subprograms.

Note that the name of the extended intrinsic subprogram must appear in a previous INTRINSIC statement in the module.

13.3.7   Return from an Invoked Module Subprogram

When the invoked subprogram executes a RETURN statement or its END FUNCTION, or END SUBROUTINE statement, the execution of the (instance of the) subprogram is terminated, control returns to the invoking main program or subprogram, and the arguments become disassociated.

If the terminated subprogram is a function, the function result is supplied for use in the expression at the place of the reference to the function.

If the terminated subprogram is a subroutine, the execution of the CALL statement in the referencing main program or subprogram is completed.

RETURN statement

The RETURN statement has the following form:

RETURN

A RETURN statement may appear only in the scoping unit of a module subprogram. A subprogram may contain more than one RETURN statement.

When a subroutine executes its END SUBROUTINE statement or when it executes a RETURN statement, the execution of the subroutine is completed and the execution of the invoking main program or subprogram continues with the first executable statement following the CALL statement.

```
module dis
  public :: discriminant
contains
  subroutine discriminant (n)
    integer, intent(inout) :: n
    select case (n)
      case (:-1)
        write (unit=*, fmt=*) " Discriminant is negative"
        return
      case (0)
        write (unit=*, fmt=*) " Discriminant is zero"
        return
      case (1:)
        write (unit=*, fmt=*) " Discriminant is positive"
    end select
    :
  end subroutine discriminant
end module dis

program ret
use dis
integer :: d
read (unit=*, fmt=*) d
call discriminant(d)
:
end program ret
```

Regardless of whether the invoked subroutine `discriminant` executes one of the RETURN statements or the END SUBROUTINE statement, control returns to statement subsequent to the CALL statement in the main program.

13.4 Internal Program Communication

Program units and subprograms may exchange information about data and subprogram names. This communication may take place via argument lists, result variables of functions, shared local variables, and external files. External files, which allow programs to communicate with the outside world, are presented in chapter 11.

13.4.1 Argument Lists

An *actual argument list* appears in a subprogram reference. A *dummy argument list* appears in a FUNCTION or SUBROUTINE statement within a subprogram definition or within an interface definition (contained in a specific interface block). The actual arguments in the subprogram reference and the corresponding dummy arguments of the subprogram become associated during the invocation of the subprogram.

The communication between these argument lists is called the **argument association**. Arguments may be used to exchange values. They may also be used to transfer particular subprogram names to the invoked subprogram.

A dummy argument is called an **input argument** if it can receive information from the invoking scoping unit but cannot transfer information back to the invoking scoping unit. A dummy argument is called an **output argument** if it can transfer information to the invoking scoping unit but cannot receive information from the invoking scoping unit. And a dummy argument is called an **input/output argument** if it can both receive information from the invoking scoping unit and transfer information to the invoking scoping unit.

Though a dummy argument is an output argument or an input/output argument, there may be circumstances such that this dummy argument is not allowed to transfer data back to the invoking scoping unit.

13.4.1.1 Dummy Argument List

Dummy arguments are used in place of actual arguments within the subprogram definition. When the subprogram is invoked, the actual arguments become associated with the corresponding dummy arguments such that they are used instead of the dummy arguments during the execution of the subprogram.

The dummy arguments of a user-defined subprogram are specified in the dummy argument list of a FUNCTION or SUBROUTINE statement. For a module subprogram this FUNCTION or SUBROUTINE statement appears in the subprogram definition. For an external subprogram or dummy subprogram, this FUNCTION or SUBROUTINE statement appears in the interface definition within the interface block for the external subprogram and dummy subprogram, respectively. The dummy arguments of *intrinsic* subprograms are predefined and need not be specified.

Characteristics of dummy arguments

The name of a dummy argument must not be specified to have an ALLOCAT-ABLE, PARAMETER, or SAVE attribute, or an initial value and must not appear in INTRINSIC statements. And the name of a dummy argument must be different from the name of the subprogram in the FUNCTION or SUBROUTINE statement.

An actual argument which is associated with a dummy argument during subprogram invocation must have certain properties. These properties depend on the characteristics of the dummy argument. Each dummy argument is either the name of a variable or the name of a subprogram.

Dummy data object: The characteristics of a dummy argument identifying a data object are: the type, the kind type parameter (if any), the character length (if any), the rank, the INTENT attribute (if any), the OPTIONAL attribute (if any), the POINTER attribute (if any), the TARGET attribute (if any), the dependence of the kind type parameter or the array bounds on the value and properties of other data entities.

Dummy subprogram: The characteristics of a dummy argument identifying a subprogram are: the classification of the dummy subprogram as a function or subroutine, the OPTIONAL attribute (if any, in case of a dummy function), the characteristics of its dummy arguments, and (in case of a dummy function) the characteristics of the function result.

13.4.1.2   Actual Argument List

Actual arguments may be specified in the actual argument list of a subprogram reference. The actual argument list may be empty. The form of the actual argument list is described in detail in connection with the description of subprogram references.

Actual arguments may be constants, variables, function references, expressions with operators and/or parentheses, and subprogram names.

Data entities: If the actual argument is a data entity, the corresponding dummy argument must be the name of a variable.

Subprogram name: If the actual argument is a subprogram name, the corresponding dummy argument must identify a dummy subprogram. Only a module subprogram name is allowed to be specified as an actual argument.

Keyword (actual) arguments

Where a subprogram is available to be referenced, the name of any dummy argument of the subprogram may be used in an actual argument list as an argument keyword.

If a keyword argument is specified in the actual argument list, all subsequent actual arguments must be specified as keyword arguments. And conversely, if an actual argument is specified as positional argument, all preceding actual arguments must be specified without an argument keyword. If there is no argument specified in the actual argument list corresponding to an optional dummy argument, an actual argument corresponding to a subsequent dummy argument may be specified only as a keyword argument.

13.4.2   Argument Association

13

During the execution of a CALL statement and during the execution of a function reference, the actual arguments (in the actual argument list of the subprogram reference) become associated with the corresponding dummy arguments (in the dummy argument list of the referenced subprogram). **Association** means here that a particular data entity or subprogram is identified both (in the scoping unit with the subprogram reference) by the actual argument and (in the referenced subprogram) by the dummy argument.

If the actual argument list does not include keyword arguments, the actual arguments and the corresponding dummy arguments become *positionally* associated; that is, the first actual argument becomes associated with the first dummy argument, the second actual argument with the second dummy argument, and so on. If a keyword argument is specified in the actual argument list, the actual argument is associated with the dummy argument which is identified by this keyword.

If there are *optional* dummy arguments in the subprogram interface, the number of actual arguments in the actual argument list of the subprogram reference may be less than or equal to the number of dummy arguments in the corresponding dummy argument list. An actual argument *must* be specified and associated for each nonoptional dummy argument, and an actual argument *may* be specified and associated for each optional dummy argument.

Normally, wherever the name of a dummy argument from the dummy argument list is used in the invoked subprogram, the corresponding actual argument is associated with this dummy argument during the subprogram invocation. This association remains in existence during the execution of the invoked subprogram, even if one or more actual arguments depend on variables which are redefined during the execution of the subprogram.

When a subprogram is recursively referenced, another instance of the subprogram is executed, which has its own set of dummy arguments. And the association of the actual arguments with the dummy arguments of this instance of the subprogram remains in existence only for the time of the execution of this instance of the subprogram.

The arguments become disassociated when a RETURN statement or the END FUNCTION or END SUBROUTINE statement of the subprogram is executed. The dummy arguments are undefined between two invocations of the same subprogram.

```
subroutine compute (fct, solution, method, strategy, print)
  interface
    function fct (x) result (ar)
      real, intent(in) :: x
      real             :: ar
    end function fct
  end interface
  real, intent(out)                 :: solution
  integer, intent(in), optional :: method, strategy, print
  :
end subroutine compute
```

Where this subroutine is available, it may be referenced as follows:

```
call compute(f, 11)
call compute(f, solution=11)
call compute(f, solution=11, method=nine, strategy=long)
call compute(f, 11, nine, strategy=long, print=2)
```

Arguments may become associated across several levels of subprogram references; that is, a dummy argument may be supplied within the same subprogram as an actual argument in a subprogram reference.

```
function fu (x)
  ⋮
  call sub(x)
  ⋮
end function fu
```

The dummy argument x is used as an actual argument in the CALL statement.

13.4.2.1 Data Objects as Dummy Arguments

If the actual argument is a data entity, the corresponding dummy argument must be a named variable, and

- The associated arguments must be of the same type;

- The associated arguments must have the same kind type parameter (if any); and

- The associated arguments must have the same rank.

Then, the associated arguments also agree with regard to shape because a non-pointer dummy array must be specified as an assumed-shape array.

Expressions, array elements, array sections, and character substrings are not associated with the corresponding dummy arguments until they are preprocessed internally to determine their values and their position, respectively. If such an actual argument includes expressions needed to determine the position, this position remains fixed for the time of the execution of the invoked subprogram, even if the expressions contain variables which are redefined during the execution of the subprogram.

If the dummy argument does not have a POINTER or TARGET attribute, any pointers associated with the actual argument do *not* become associated with the dummy argument. They remain associated with the actual argument. If the dummy argument is not a pointer and the corresponding actual argument is a pointer, the actual argument must be currently associated with a target and the dummy argument becomes argument associated with that target. If the dummy argument has the TARGET attribute and the corresponding actual argument has no TARGET attribute or is an array section with a vector subscript, pointers associated with the dummy argument become undefined when control returns to the invoking scoping unit. If the dummy argument and the associated actual argument have the TARGET attribute and if the actual argument is not an array section with a vector subscript, then the pointers associated with the actual argument become associated with the corresponding dummy argument; and after termination of the subprogram, the pointers being associated with the dummy argument remain associated with the actual argument.

Derived type arguments

An actual argument of a derived type may be associated with a dummy argument of the same type. This is guaranteed if the *same* derived type definition is used to declare the type of both the actual argument and the corresponding dummy argument. This is only possible if the derived type definition is available by host association or by USE association in the different scoping units of the arguments.

13.4.2.2 Implicit Association of Two Dummy Arguments

The reference to a subprogram may cause one dummy argument of the referenced subprogram to become associated with another dummy argument of the referenced subprogram.

```
call sub(a, a)
⋮

subroutine sub (x, y)
  ⋮
end subroutine sub
```

The same actual argument **a** is associated with both dummy arguments. Thus these dummy arguments become implicitly associated with each other.

Such a subprogram invocation must not cause one of the implicitly associated dummy arguments to be defined, redefined, or undefined during the execution of the invoked subprogram (including any other subprogram invoked from the first subprogram).

If two actual arguments partially overlap, those portions of the corresponding dummy arguments which correspond to the overlapped portions of the associated actual arguments must not be defined, redefined, or become undefined during the execution of the subprogram.

13.4.2.3 Length of Character Dummy Arguments

If the dummy argument is of type character, the associated actual argument also must be of type character. The character length of the dummy argument must be specified as an *. That's why such a dummy argument automatically assumes its length from the associated actual argument during the invocation of the subprogram.

```
module mod
  public :: sub, chrfct
contains
  subroutine sub ()
    character (len=6), save :: temp = "milky "
    character (len=10)       :: z
    z = chrfct(temp)                ! <-- 2nd invocation
    :
  end subroutine sub

  function chrfct (darg) result (chain)
    character (len=*)  :: darg
    character (len=10) :: chain
    :
    chain = darg // "way "
  end function chrfct
end module mod

program abc
use mod
character (len=4)  :: actarg
character (len=15) :: y
actarg = "rail"
y = chrfct(actarg) // "station"  ! <-- 1st invocation
:
end program abc
```

Dummy argument **darg** in function **chrfct** has length 4 during the 1st invocation and length 6 during the 2nd invocation of the function.

13.4.2.4   Scalar Arguments

A scalar dummy argument may become associated with a scalar actual argument at most. If an actual argument is scalar, the corresponding dummy argument also must be scalar.

13.4.2.5   Dummy (Argument) Arrays

A dummy argument being an array is called a **dummy argument array** or **dummy array**. The associated actual argument must be an array of the same rank.

Assumed-Shape Arrays

If the dummy argument is a nonpointer array, it must be specified to be an assumed-shape array. Thus, each dimension of the dummy argument array automatically has the same extent as the corresponding dimension of the associated actual argument array.

If a dummy argument and its associated actual argument are array entities, the array elements of the actual argument become associated with the array elements of the dummy argument in array element order. This way, corresponding array elements (with the same position) in the actual argument array and in the dummy argument array become associated with each other.

Variable dummy arrays

If an array bound specification for a dummy array is a nonconstant expression, the array is called a **variable (dummy) array**.

The lower bounds may be specification expressions that include variables and function references. The upper bounds must not be specified. The variables may be scalar integer dummy arguments or scalar integer variables which are accessible by host association or by USE association. The referenced functions must be scalar integer intrinsic functions and their actual arguments must also be such specification expressions as described here.

```
function xmean (l, x) result (res)
  integer, intent(in)            :: l
  real, intent(in), dimension (l:) :: x
  real                           :: res
  integer :: n
  n = size(x)
  res = 0.0                ! \
  do i=l,l+n-1             ! |
    res = res + x(i)       ! >  res = sum(x) / n
  enddo                    ! |
  res = res / n            ! /
end function xmean
```

The lower bound of the dummy array x is transferred to the function via the other dummy argument l. The size of the dummy array is determined within the function by a reference to the intrinsic SIZE function.

13.4.2.6   Dummy (Argument) Pointers

If the dummy argument is a pointer, the associated actual argument also must be a pointer.

During argument association, the dummy argument receives the pointer association status of the actual argument. If the actual argument is currently pointer-associated, the dummy argument pointer becomes associated with the target which is associated with the actual argument. The association with this target may be modified within the subprogram by execution of an ALLOCATE statement or pointer assignment statement. In such a case, the actual argument receives the pointer association status of the dummy argument at termination of the subprogram, unless the target of the dummy argument is not a variable that becomes undefined when control returns from the subprogram; if the target of the dummy argument becomes undefined at termination of the subprogram, the actual argument gets the association status "undefined".

A pointer with an undefined association status must not be reused until it has got a defined association status by the execution of an ALLOCATE statement, pointer assignment statement, or NULLIFY statement.

13.4.2.7   Restrictions on the Association of Data Entities

For the time of the execution of a subprogram, there are certain restrictions on arguments which are data entities. Such a dummy argument must *not* be redefined

13

- If the associated actual argument is a constant, a function reference, or another expression with operator(s) or enclosed in parentheses;

- If the associated actual argument appears more than once in the actual argument list; or

- If the associated actual argument is an array section with a vector subscript.

While a data entity is associated with a dummy argument, its allocation status must not be modified. While a data entity is associated with a dummy argument, any action, which affects the value of the data entity or of a part of it, must be taken only through the dummy argument, except the dummy argument has the POINTER attribute, the part is a pointer object or a part of a pointer object, or the dummy argument has the TARGET attribute but no INTENT(IN) attribute, or the dummy argument is no pointer and the actual argument is a target that is no array section with a vector subscript. If any part of the data entity is defined as a result of a definition of the dummy argument, this part must be referenced during the execution of the subprogram only through the dummy argument; this rule applies to the whole scoping unit of the subprogram.

13.4.2.8 Dummy Subprograms

Only the specific name of a module subprogram may be specified as an actual argument corresponding to a dummy subprogram. A dummy subprogram of a function also must be a function.

If a specific name which is supplied as an actual argument is equal to a generic name, only the specific name is associated; i. e., the dummy argument does not receive the generic properties (if any) of the actual argument.

13.4.3 Optional Dummy Arguments

A dummy argument which has the OPTIONAL attribute is an **optional** dummy argument. If a dummy argument has the OPTIONAL attribute all subsequent dummy arguments also must have the OPTIONAL attribute. In the subprogram reference, no actual argument need be specified corresponding to an optional dummy argument; in this case, no actual argument is associated with the optional dummy argument during the execution of the subprogram.

But if an actual argument is specified corresponding to an optional dummy argument, this actual argument may itself be a dummy argument of the subprogram which contains the subprogram reference. If this dummy argument of the outer subprogram is also an optional argument, no actual argument need be specified in the reference to this subprogram. But if an actual argument is specified corresponding to this optional dummy argument, this actual argument may itself be a dummy argument of the subprogram which contains the subprogram reference. — And so on, for any depth of nested references to subprograms with optional dummy arguments.

An optional dummy argument is **not present** during subprogram execution if no actual argument is associated with it. The following restrictions apply to an optional dummy argument that is not present:

- If it is a dummy data object, it must not be referenced or defined.

- If it is a dummy subprogram, it must not be invoked during program execution.

- It must not be specified as an actual argument corresponding to a nonoptional dummy argument except in a reference to the PRESENT intrinsic function.

- A subobject of it must not be specified as an actual argument corresponding to an optional dummy argument.

- If it is a pointer, it must not be specified as an actual argument corresponding to a nonpointer dummy argument except in a reference to the PRESENT intrinsic function.

Except as noted in the above list, an optional dummy argument which is not present may be specified as an actual argument corresponding to an optional dummy argument, and this optional dummy argument is then also considered to be not associated with an actual argument. That is, the property of presence may be inherited from the actual argument.

A dummy argument is **present** during subprogram execution if an actual argument is associated with it which is no dummy argument in the referencing scoping unit. A present dummy argument may be specified as an actual argument corresponding to a dummy argument, then this last dummy argument also is considered to be present. A nonoptional dummy argument must be present.

13.4.4 Dummy Argument with INTENT Attribute

A dummy argument which is a nonpointer data object must be specified to have an INTENT attribute specifying the intended use of the dummy argument.

A dummy argument with INTENT(IN) attribute is an *input argument* which may be used but must not be defined or become undefined within the subprogram. A dummy argument with INTENT(OUT) attribute is an *output argument*. Such a dummy argument is initially undefined during subprogram execution. And a dummy argument with INTENT(INOUT) attribute is an *input/output argument*.

A dummy argument with INTENT(IN) or a subobject of such a dummy argument must not appear as the left-hand side of an assignment statement, as an input list item in a READ statement, as an internal file in a WRITE statement, as the integer variable in an IOSTAT= or SIZE= specifier in an input/output statement, as a definable variable in a specifier in an INQUIRE statement, as a status variable in an ALLOCATE or DEALLOCATE statement, or as an actual argument in a reference to a subprogram that is able to redefine the actual argument.

If the INTENT(OUT) or INTENT(INOUT) attribute is specified for a dummy argument, it must be definable. If the INTENT(OUT) attribute is specified for a dummy argument, the actual argument automatically becomes undefined when it becomes associated with the dummy argument.

14 INTRINSIC SUBPROGRAMS

An intrinsic subprogram is a predefined subprogram supplied by the F processor. An F processor must supply at least the intrinsic subprogram that are described in this chapter.

14.1 Intrinsic Functions

Intrinsic functions may be classified as *inquiry functions*, *elemental functions*, or *transformational functions*.

Inquiry functions return information about those properties of the argument(s) that do not depend on the value(s) of the argument(s). If an actual argument inquired about consists of a single variable, it may be undefined.

Elemental functions are defined for scalar arguments, but they also can process array arguments element-by-element in order to return a corresponding array result.

Transformational functions are all other intrinsic functions. Nearly all of them are defined for array arguments. The rank of the function result may be different from the rank of the argument(s).

Most of the intrinsic functions may be optionally referenced with actual arguments of different datatypes. Then, except for a reference to the INT, REAL, NINT, or ABS function, the result automatically has the same datatype as the first actual argument.

```
integer, parameter :: double = selected_real_kind(14,200)
print *, sqrt(2.0)
print *, sqrt(2.0_double)
```

The first PRINT statement prints the value $\sqrt{2}$ as a (default) real value, such as 1.414214; and the second PRINT statement prints $\sqrt{2}$ as a real value with more precision, such as 1.414213562373096.

The name of an intrinsic function must not be supplied as an actual argument in a subprogram reference.

Masked actual arguments: The function definitions of some intrinsic functions contain an *optional* argument MASK, which may be used to specify a logical mask. This logical mask is useful if not all but only selected array elements of another array argument are to be processed by the function. If this optional argument MASK is present, those array elements of the argument need *not* be defined which are *not* selected by the mask. Note that some intrinsic functions contain a *nonoptional* argument MASK which also defines a logical mask.

14.1.1   Table of Intrinsic Functions

Name	No. of arg.	Type of 1st arg.	Type of result	Description
ABS	1	integer	integer	Absolute value
		real	real	$y = |x|$
		complex	real	$y = \sqrt{(\Re(x))^2 + ((\Im(x))^2}$
ACOS	1	real	real	Arccosine: $y = \arccos x$
				y in radians
ADJUSTL	1	character	character	Adjust left
ADJUSTR	1	character	character	Adjust right
AIMAG	1	complex	real	Imaginary part of complex
				argument: $y = \Im(x)$
AINT	1 [,2]	real	real	Truncate: $y = \text{int } x$
ALL	1 [,2]	logical	logical	True if all values true
ALLOCATED	1	any	logical	Allocation status
ANINT	1 [,2]	real	real	Nearest whole number:
				$y = \text{int}(x + 0.5),\ x \geq 0$
				$y = \text{int}(x - 0.5),\ x < 0$
ANY	1 [,2]	logical	logical	True if one value true
ASIN	1	real	real	Arcsine: $y = \arcsin x$
				y in radians
ASSOCIATED	1 [,2]	any	logical	Pointer association status
ATAN	1	real	real	Arctangent: $y = \arctan x$
				y in radians
ATAN2	2	real	real	Arctangent: $y = \arctan x1/x2$
				y in radians
BIT_SIZE	1	integer	integer	Number of Bits
BTEST	2	integer	logical	Tests a bit
CEILING	1 [,2]	real	integer	Smallest integer $\geq x$
CHAR	1 [,2]	integer	character	Type conversion:
				integer to character
CMPLX	1 [,2,3]	integer	complex	Type conversion:
		real	complex	numeric to complex
		complex	complex	
CONJG	1	complex	complex	Conjugate: $y = \Re(x) - i\,\Im(x)$

Name	No. of arg.	Type of 1st arg.	Type of result	Description
COS	1	real	real	Cosine: $y = \cos x$
		complex	complex	x in radians
COSH	1	real	real	Hyperbolic cosine: $y = \cosh x$
COUNT	1 [,2]	logical	integer	Number of true elements
CSHIFT	3	any	as 1. arg.	Circular shift
DIGITS	1	integer	integer	Number of significant digits
		real	integer	in the model of x
DOT_ PRODUCT	2	numeric	numeric	Dot product
		logical	logical	
EOSHIFT	3 [,4]	any	as 1. arg.	End-off shift
EPSILON	1	real	real	Smallest number with
				exponent 0
EXP	1	real	real	Exponential function:
		complex	complex	$y = e^x$
EXPONENT	1	real	integer	Exponent part of x in model
FLOOR	1 [,2]	real	integer	Largest integer $\leq x$
FRACTION	1	real	real	Fractional part of x in
				the model
HUGE	1	integer	integer	Largest integer in the model
		real	real	
IAND	2	integer	integer	Logical AND
IBCLR	2	integer	integer	Clear bit
IBITS	3	integer	integer	Bit extraction
IBSET	2	integer	integer	Set bit
ICHAR	1	character	integer	Type conversion:
				character to integer
IEOR	2	integer	integer	Exclusive OR
INDEX	2 [,3]	character	integer	Index of a substring in
				character string
INT	1 [,2]	integer	integer	Type conversion:
		real	integer	numeric to integer
		complex	integer	
IOR	2	integer	integer	Inclusive OR
ISHFT	2	integer	integer	Logical shift
ISHFTC	2 [,3]	integer	integer	Circular shift

14

Name	No. of arg.	Type of 1st arg.	Type of result	Description
KIND	1	any	integer	Kind type parameter
LBOUND	1 [,2]	any	integer	Lower array bounds
LEN	1	character	integer	Length of character string
LEN_TRIM	1	character	integer	Length without trailing blanks
LOG	1	real complex	real complex	Natural logarithm: $y = \ln x$, base e
LOGICAL	1 [,2]	logical	logical	Kind param. conversion
LOG10	1	real	real	Common logarithm: $y = \log x$, Base 10
MATMUL	2	numeric logical	numeric logical	Matrix multiplication
MAX	2 [,3,...]	integer real	integer real	Largest value: $y = \max(x1, x2, ...)$
MAXEXPONENT	1	real	integer	Largest exponent in model
MAXLOC	1 [,2]	integer real	integer integer	Location of largest value of array
MAXVAL	1 [,2,3]	integer real	integer real	Largest value of array
MERGE	3	any	as 1. arg.	Merge 2 arrays
MIN	2 [,3,...]	integer real	integer real	Smallest value: $y = \min(x1, x2, ...)$
MINEXPONENT	1	real	integer	Smallest exponent in model
MINLOC	1 [,2]	integer real	integer integer	Location of smallest value of array
MINVAL	1 [,2,3]	integer real	integer real	Smallest value of array
MODULO	2	integer real	integer real	Modulo function
NEAREST	2	real	real	Nearest different representable number
NINT	1 [,2]	real	integer	Nearest integer: $y = \mathrm{int}(x + 0.5),\ x \geq 0$ $y = \mathrm{int}(x - 0.5),\ x < 0$
NOT	1	integer	integer	Logical complement

Name	No. of arg.	Type of 1st arg.	Type of result	Description
PACK	2 [,3]	any	as 1. arg.	Pack array 1-dimensional
PRECISION	1	real	integer	Decimal precision
		complex	integer	
PRESENT	1	any	logical	Argument presence
PRODUCT	1 [,2,3]	numeric	as 1. arg.	Product of array elements
RADIX	1	integer	integer	Base of model
		real	integer	
RANGE	1	numeric	integer	Decimal exponent range
REAL	1 [,2]	integer	real	Type conversion:
		real	real	numeric to real
		complex	real	
REPEAT	2	character	character	Repeated concatenation
RESHAPE	2 [,3,4]	integer	as 2. arg.	Reshape array
RRSPACING	1	real	real	Reciprocal of rel. spacing of numbers near x
SCALE	2	real	real	Scaling: $y = x1 * b^{x2}$
SCAN	2 [,3]	character	integer	Position of a character
SELECTED_ INT_KIND	1	integer	integer	Smallest kind type parameter of an int. type
SELECTED_ REAL_KIND	1 [,2]	integer	integer	Smallest kind type parameter of a real type
SET_ EXPONENT	2	real	real	Set exponent part of a number
SHAPE	1	any	integer	Shape of array/scalar
SIGN	2	integer	integer	Transfer of sign:
		real	real	$y = \lvert x1 \rvert, \ x2 \geq 0$ $y = -\lvert x1 \rvert, \ x2 < 0$
SIN	1	real	real	Sine: $y = \sin x$
		complex	complex	x in radians
SINH	1	real	real	Hyperbolic sine: $y = \sinh x$
SIZE	1 [,2]	any	integer	Number of array elements
SPACING	1	real	real	Absolute spacing near x
SPREAD	3	any	as 1. arg.	Adding a dimension to array
SQRT	1	real	real	Square root: $y = \sqrt{x}$
		complex	complex	

14

Name	No. of arg.	Type of 1st arg.	Type of result	Description
SUM	1 [,2,3]	numeric	as 1. arg.	Sum of array elements
TAN	1	real	real	Tangent: $y = \tan x$ x in radians
TANH	1	real	real	Hyperbolic tangent: $y = \tanh x$
TINY	1	real	real	Smallest number in the model
TRANSPOSE	1	any	as arg.	Transpose of array
TRIM	1	character	character	Remove trailing blanks
UBOUND	1 [,2]	any	integer	Upper array bounds
UNPACK	3	any	as 1. arg.	Unpack 1-dim. array
VERIFY	2 [,3]	character	integer	Verify set of characters in a string

14.2 Intrinsic Subroutines

Name	No. of arg.	Description
DATE_AND_TIME	1 [,2,3,4]	Returns real time clock and date
MVBITS	5	Copies bits between integer objects
RANDOM_NUMBER	1	Returns pseudorandom number
RANDOM_SEED	0 [,1]	Queries or (re)starts pseudorandom number generator
SYSTEM_CLOCK	1 [,2,3]	Returns real-time clock information

The name of an intrinsic subroutine must *not* be specified as an actual argument in a subprogram reference.

14.3 Intrinsic Subprogram Reference

A reference to an intrinsic subprogram has the same form as a reference to a user-defined subprogram. On invocation, the actual arguments in the reference to an intrinsic subprogram become associated with the corresponding dummy arguments in the subprogram definition.

Usually, the type of the function result is determined by the type of the first actual argument in the function reference.

Keyword (actual) arguments: In any scoping unit of a program, actual arguments in references to intrinsic subprograms may be specified as keyword arguments. The specification of a keyword (actual) argument in a reference to an intrinsic subprogram must satisfy the same rules as a keyword argument in a reference to a user-defined subprogram.

```
i = index(z, substring="kap.")
y = exp(x=ee)
```

A kind type parameter value in a literal constant which is supplied as an actual argument in a reference to an intrinsic subprogram must be written as a named constant.

KIND dummy argument: An actual argument associated with the intrinsic function dummy argument KIND must be a scalar integer named constant with a result that specifies a representation method that is supported by the F processor.

Elemental intrinsic subprogram reference: When an elemental function is referenced, the function result has the same shape as the actual argument with the largest rank. The function reference may include array arguments. If it includes more than one array argument, all array arguments must have the same shape. Then, the function result also is an array and has the same shape as the array argument(s). It is obtained as if each of its elements is calculated separately by evaluating the function using the corresponding element(s) and the scalar argument(s). The order of these elemental evaluations has no effect on the function result.

An optional array argument that is not present must not be supplied as an actual argument in a reference to an elemental intrinsic function unless an array of the same rank is supplied as an actual argument corresponding to a nonoptional dummy argument of that elemental function.

14

A reference to the MVBITS elemental subroutine may include an array argument corresponding to its dummy argument TO. In this case, the remaining actual arguments must be conformable with this actual argument. And the values of the elements of this actual argument array are the same as would be obtained if MVBITS were applied separately to corresponding elements of each argument. In a reference to the intrinsic subroutine MVBITS, the actual arguments corresponding to the TO and FROM dummy arguments may be the same variable.

Inquiry functions: If the actual argument being inquired about in a reference to BIT_SIZE, DIGITS, EPSILON, HUGE, MAXEXPONENT, MINEXPONENT, PRECISION, RADIX, RANGE, or TINY is a pointer, it may have undefined or disassociated association status, and if it is allocatable, it may not be allocated.

14.4   Intrinsic Subprogram Definitions

The following are the descriptions of the intrinsic functions and intrinsic subroutines. Optional arguments appearing in headings and tables within this book are enclosed in square brackets. This bracket notation cannot describe all legal forms for argument usage. Note that the actual argument list in a reference to an intrinsic subprogram must not contain a leading comma, and that there are some intrinsic subprograms with more than one optional argument requiring at least one of its optional arguments to be present.

Note that the nonpointer dummy arguments of the intrinsic functions have INTENT(IN) if another intent is not stated explicitly.

ABS (A)

ABS is an elemental function, which returns the absolute value of the argument. The argument A must be of numeric type. If A is of type integer or real, the result is of the same type and kind type parameter as A. If A is of type complex, the result is of type real and is a processor-dependent approximation to $\sqrt{(\Re(A))^2 + (\Im(A))^2}$.

ACOS (X)

ACOS is an elemental function, which returns the arccosine expressed in radians. The argument X must be of type real with $|X| \leq 1$. The result is of type real with the same kind type parameter as X. The range of the result is $0 \leq ACOS(X) \leq \pi$.

ADJUSTL (STRING)

ADJUSTL is an elemental function, which adjusts a given character string left by removing all leading blanks and inserting the same number of trailing blanks. The argument STRING must be of type character. The result is of type character with the same length as STRING.

ADJUSTR (STRING)

ADJUSTR is an elemental function, which adjusts a given character string right by removing all trailing blanks and inserting the same number of leading blanks. The argument STRING must be of type character. The result is of type character with the same length as STRING.

AIMAG (Z)

AIMAG is an elemental function, which returns the imaginary part of the complex argument Z. The result is of type real and has the same kind type parameter as the argument Z.

AINT (A [, KIND])

AINT is an elemental function, which truncates the argument towards zero. The argument A must be of type real. The result is of type real. If KIND is present, the result has the kind type parameter KIND; otherwise it is of type default real.

ALL (MASK [, DIM])

ALL is a transformational function, which returns the value *true* if all elements in the argument MASK are *true*, or if DIM is present, if all elements in MASK along dimension DIM are *true*. The argument MASK must be an array of type logical. The optional argument DIM must be scalar and of type integer. The result is of type logical and has the same kind type parameter as MASK.

ALLOCATED (ARRAY)

ALLOCATED is an inquiry function, which returns the value *true* if the argument ARRAY is currently allocated. The argument ARRAY must be an allocatable array (of any type). The result is scalar of type default logical.

ANINT (A [, KIND])

ANINT is an elemental function, which returns the nearest whole number to the argument A. The argument A must be of type real. If KIND is present, the result is of type real with the kind type parameter KIND; otherwise it has the same kind type parameter as the argument A.

ANY (MASK [, DIM])

ANY is a transformational function, which returns the value *true* if any of the elements in the argument MASK are *true*, or if DIM is present, if any of the elements in MASK along dimension DIM are *true*. The argument MASK must be an array of type logical. The optional argument DIM must be scalar and of type integer. The result is of type logical and has the same kind type parameter as MASK.

14

ASIN (X)

ASIN is an elemental function, which returns the arcsine expressed in radians. The argument X must be of type real with $|X| \leq 1$. The result is of type real with the same kind type parameter as X. The range of the result is $-\pi/2 \leq ASIN(X) \leq \pi/2$.

ASSOCIATED (POINTER [, TARGET])

ASSOCIATED is an inquiry function, which indicates whether the given POINTER is currently associated with a target or whether the given POINTER is currently associated with the given TARGET. The argument POINTER must be a pointer (of any type); its pointer association status must not be undefined. The optional argument TARGET must be a pointer or a target. It must have the same type, kind type parameter (if applicable), character length (if applicable) and rank as POINTER. The result is scalar and of type default logical.

If TARGET is not present, the result is *true* if POINTER is currently associated with a target; otherwise the result is *false*.

If TARGET is present and is a target, the result is *true* if POINTER is associated with TARGET; otherwise the result is *false*.

If TARGET is present and is a pointer, the result is *true* if the target associated with POINTER is the same target which is associated with TARGET; otherwise or if either POINTER or TARGET is disassociated, the result is *false*.

ATAN (X)

ATAN is an elemental function, which returns the arctangent expressed in radians. The argument X must be of type real. The result is of type real with the same kind type parameter as X. The range of the result is $-\pi/2 \le \text{ATAN}(X) \le \pi/2$.

ATAN2 (Y, X)

ATAN2 is an elemental function, which returns the arctangent for a pair of arguments. The arguments must be of type real; their kind type parameters must agree. The result is of type real with the same kind type parameter as the arguments. The result is a processor-dependent approximation to the principal value of the complex number (X, Y) and is expressed in radians. If $X \neq 0$, the result is $\arctan Y/X$, and the range of the result is $-\pi < \text{ATAN2}(Y, X) \le \pi$. The arguments Y and X must not both be zero at the same time.

BIT_SIZE (I)

BIT_SIZE is an inquiry function, which returns the number of bits s in the model for bits within an integer of the same kind type parameter as I. The argument I may be scalar or array valued and must be of type integer. If the actual argument for I is a variable name, the variable need not be defined, if it is a pointer its association status is irrelevant, and if it is an allocatable array its allocation status is irrelevant. The result is a scalar integer with the same kind type parameter as I.

BTEST (I, POS)

BTEST is an elemental function, which tests a bit in the integer argument I. POS must be of type integer with $0 \leq POS < BIT\_SIZE(I)$. The result is of type default logical. The result has the value *true* if bit POS in I has the value 1, and it has the value *false* if bit POS in I has the value 0.

CEILING (A)

CEILING is an elemental function, which returns the least default integer greater than or equal to its argument. The argument A must be of type real. The result is of type default integer.

CHAR (I)

CHAR is an elemental function, which returns the character in position I in the ASCII collating sequence. The argument I must be of type integer with $0 \leq I \leq 127$. The result is of type character with length 1. Note:

$ICHAR(CHAR(i)) = i$ for $0 \leq i \leq 127$.
$CHAR(ICHAR(z)) = z$ for any character z
                             which the F processor can represent.

CMPLX (X [, Y] [, KIND])

CMPLX is an elemental function, which converts X or (X, Y), respectively, to complex type. The argument X must be of numeric type. The optional argument Y must be of type integer or real. The result is of type complex. If the argument KIND is present, the result has the kind type parameter KIND; otherwise it is of type default complex.

If Y is absent and X is not of type complex, the result has the same value as CMPLX(X, 0) or CMPLX(X, 0, KIND), respectively.

If Y is absent and X is of type complex, the result has the same value as CMPLX(X, AIMAG(X)) or CMPLX(X, AIMAG(X), KIND), respectively.

CMPLX(X, Y, KIND) has the complex value
(REAL(X, KIND), REAL(Y, KIND)).

CONJG (Z)

CONJG is an elemental function, which returns the conjugate of the complex argument Z. The result is of type complex with the same kind type parameter as Z.

COS (X)

COS is an elemental function, which returns the cosine. The argument X must be of type real or complex and must be specified in radians. The result is of the same type and kind type parameter as X.

COSH (X)

COSH is an elemental function, which returns the hyperbolic cosine. The argument X must be of type real. The result is of type real with the same kind type parameter as X.

COUNT (MASK [, DIM])

COUNT is a transformational function, which returns the number of elements in the logical array MASK that have the value *true*, or if DIM is present, which returns the number of *true* elements in MASK along dimension DIM. The optional argument DIM must be a scalar of type integer. The result is of type default integer.

CSHIFT (ARRAY, SHIFT [, DIM])

CSHIFT is a tranformational function, which performs a circular shift on the 1-dimensional array ARRAY or performs a circular shifts on all complete 1-dimensional sections along a given dimension of the 2- or multi-dimensional array ARRAY. Array elements shifted out at one end of an array section are shifted in at the other end. Different array sections may be shifted by different amounts and in different directions. The argument ARRAY must be an array (of any type). The optional argument DIM must be a scalar of type integer. If DIM is omitted, the default DIM=1 is used. The argument SHIFT must be of type integer, it must be scalar if ARRAY is a 1-dimensional array. The result is of the same type, kind type parameter (if applicable), and character length (if applicable) as ARRAY. If SHIFT is scalar, the result is obtained by shifting every 1-dimensional array section that extends across the dimension DIM circularly SHIFT times. If SHIFT=1 and ARRAY is 1-dimensional, the array is shifted circularly to the left such that the first element of ARRAY becomes the last element of the result. If SHIFT is an array, it must have the shape of ARRAY with dimension DIM omitted. In this case, SHIFT must supply a separate value for each shift.

DATE_AND_TIME ([DATE] [, TIME] [, ZONE] [, VALUES])

DATE_AND_TIME is an intrinsic subroutine, which returns through its output arguments information from the real-time clock. The form of the date and time information is compatible with the representations defined in ISO 8601:1988.

DATE is an optional INTENT(OUT) argument and must be a scalar variable of type character and of length ≥ 8. After return from the subroutine, the leftmost 8 characters contain a string of the form *CCYYMMDD*, with *CC* being the century, *YY* the year within the century, *MM* the month within the year, and *DD* the day within the month. If the F processor does not support a calendar, DATE is set to blank.

TIME is an optional INTENT(OUT) argument and must be a scalar variable of type character and of length ≥ 10. After return from the subroutine, the leftmost 10 characters contain a string of the form *hhmmss.sss*, with *hh* being the hour of the day, *mm* the minutes of the hour, and *ss.sss* the seconds and the milliseconds of the minute. If the F processor does not support a clock, TIME is set to blank.

ZONE is an optional INTENT(OUT) argument and must be a scalar variable of type character and of length ≥ 5. After return from the subroutine, the leftmost 5 characters contain a string of the form $\pm$*hhmm*, where *hh* and *mm* are the time difference with respect to UTC. If the F processor does not support a clock, ZONE is set ot blank.

VALUES is an optional INTENT(OUT) argument and must be a 1-dimensional array variable of type default integer and size ≥ 8. After return from the subroutine, the first 8 elements contain the following values: the year (for example, 1996), the month of the year, the day of the month, the time difference with respect to UTC in minutes, the hour of the day in the range of 0 to 23, the minutes of the hour in the range of 0 to 59, the seconds of the minute in the range of 0 to 60, and the milliseconds of the second in the range of 0 to 999. If the F processor does not support a calendar or a clock, the corresponding elements of VALUES are set to $-$HUGE(0).

14

DIGITS (X)

DIGITS is an inquiry function, which returns the number of significant digits in the model that includes the argument X; this is q if X is of type integer; otherwise it is p (see appendix B). X must be a scalar or an array of type integer or real. The result is of type default integer.

DOT_PRODUCT (VECTOR_A, VECTOR_B)

DOT_PRODUCT is a transformational function, which performs dot-product multiplication of two numeric or two logical 1-dimensional arrays. The argument arrays must have the same size. If VECTOR_A is of numeric type, VECTOR_B may be of the same or any other numeric type. If the arguments are of numeric type, the type and kind type parameter of the result are those of the expression (VECTOR_A * VECTOR_B). If VECTOR_A is of type integer or real, the result is scalar and has the value SUM(VECTOR_A * VECTOR_B).

And if VECTOR_A is of type complex, the result is scalar and has the value SUM(CONJG(VECTOR_A) * VECTOR_B). If the arguments are of type logical, the result also is of type logical; the kind type parameter is that of (VECTOR_A .AND. VECTOR_B). And the result is scalar and has the value ANY(VECTOR_A .AND. VECTOR_B).

EOSHIFT (ARRAY, SHIFT [, BOUNDARY] [, DIM])

EOSHIFT is a transformational function, which returns a result that is an end-off shift on the 1-dimensional array ARRAY or is an end-off shift on all complete 1-dimensional sections along a given dimension of the 2- or multi-dimensional array ARRAY. Array elements are shifted out at one end of a section and copies of a boundary value BOUNDARY are shifted in at the other end. BOUNDARY may be omitted for intrinsic types of ARRAY, in which case the default boundary value of zero, *false*, or blank, respectively, is inserted. BOUNDARY must be of the same type and (if applicable) kind type parameter as ARRAY. The optional argument DIM must be scalar and of type integer. Different sections may have different boundary values and may be shifted by different amounts of SHIFT and in different directions. ARRAY must be an array of any type. The argument SHIFT must be of type integer. SHIFT must be scalar, if ARRAY is 1-dimensional; otherwise SHIFT may be scalar, or SHIFT may be an array whose shape is that of ARRAY with dimension DIM omitted, and must supply a separate value for each shift. The result is of the same type, (if applicable) kind type parameter, and (if applicable) character length as ARRAY.

EPSILON (X)

EPSILON is an inquiry function, which returns a positive model number that is almost negligible compared with the value one in the model that includes X. The argument must be of type real. The result is scalar and is of type real with the same kind type parameter as X. The result has the value b^{1-p} (see app. B).

EXP (X)

EXP is an elemental function, which returns the value of the exponential function e^X. The argument X must be of type real or complex. The result is of the same type and kind type parameter as X.

EXPONENT (X)

EXPONENT is an elemental function, which returns the exponent part e of the argument X when represented as a model number. The argument X must be of type real. The result is of type default integer. If $X = 0$, the result has the value zero.

FLOOR (A)

FLOOR is an elemental function, which returns the greatest integer less than or equal to the argument. The argument must be of type real. The result is of type default integer.

FRACTION (X)

FRACTION is an elemental function, which returns the fractional part of the model representation of the argument. The argument X must be of type real. The result is of type real with the same kind type parameter as X. The result has the value $X \times b^{-e}$ (see appendix B).

HUGE (X)

HUGE is an inquiry function, which returns the largest number in the model that includes X. The argument X may be a scalar or an array and must be of type integer or real. The result is scalar and is of the same type and kind type parameter as X. If X is of type integer, the result has the value $(r^q - 1)$; and if X is of type real, the result has the value $(1 - b^{-p})b^{e_{max}}$ (see appendix B).

IAND (I, J)

IAND is an elemental function, which returns the logical AND of all bits in I and corresponding bits in J. The arguments I and J must be of type integer with the same kind type parameter. The result is of type integer and has the same kind type parameter as the arguments. Integers which are sequences of bits are interpreted according to the model for bit manipulation. The following truth table shows the effect of the function:

I	1	1	0	0
J	1	0	1	0
IAND(I, J)	1	0	0	0

IBCLR (I, POS)

IBCLR is an elemental function, which returns an integer with the same sequence of bits as the argument I except that bit position POS is set to 0. The argument I must be of type integer and is interpreted according to the model for bit manipulation as a sequence of bits. POS must be of type integer with $0 \le POS < BIT\_SIZE(I)$. The result is of the same type and kind type parameter as I.

IBITS (I, POS, LEN)

IBITS is an elemental function, which returns a subsequence of the given sequence of bits in I. The argument I must be of type integer. POS must be of type integer with $POS \geq 0$ and $POS + LEN \leq BIT\_SIZE(I)$. LEN must be a nonnegative integer. The result is of type integer with the same kind type parameter as I. The result contains right adjusted the same sequence of bits as the LEN bits of I starting at bit POS; any remaining leading bits of the result are set to 0. Integers which are sequences of bits are interpreted according to the model for bit manipulation.

IBSET (I, POS)

IBSET is an elemental function, which returns an integer with the same sequence of bits as the argument I except that bit position POS is set to 1. The argument I must be of type integer and is interpreted according to the model for bit manipulation as a sequence of bits. POS must be of type integer with $0 \leq POS < BIT\_SIZE(I)$. The result is of the same type and kind type parameter as I.

ICHAR (C)

ICHAR is a function, which returns the position of a given character in the ASCII collating sequence. The argument is of type character and of length 1. The result is of type default integer. Note:
$ICHAR(CHAR(i)) = i$ and $CHAR(ICHAR(z)) = z$.

IEOR (I, J)

IEOR is an elemental function, which returns the logical exclusive OR of all bits in I and corresponding bits in J. The arguments I and J must be of type integer with the same kind type parameter. The result is of type integer and has the same kind type parameter as the arguments. Integers which are sequences of bits are interpreted according to the model for bit manipulation. The following truth table shows the effect of the function:

I	1	1	0	0
J	1	0	1	0
IEOR(I, J)	0	1	1	0

INDEX (STRING, SUBSTRING [, BACK])

INDEX is an elemental function, which returns the starting position of SUBSTRING as a substring of STRING. The arguments STRING and SUBSTRING must be of type character. The optional argument BACK must be of type logical. The result is of type default integer. The result is 0 if the substring does not occur in STRING or if the length of STRING is less than the length of SUBSTRING. If BACK is absent or present with the value *false*, the result is the position of the first occurrence of the substring SUBSTRING; if the length of SUBSTRING is zero, the result is 1. If BACK is present with the value *true*, the result is the position of the last occurrence of the substring SUBSTRING; if the length of SUBSTRING is zero, the result is LEN(STRING) + 1.

INT (A [, KIND])

INT is an elemental function, which performs a type conversion. The argument A must be of numeric type. If KIND is present, the result is of type integer and has the kind type parameter KIND; otherwise the result is of type default integer. If A is of type integer, the result has the value A. If A is of type real, the result is A truncated towards zero. And if A is of type complex, the result is the real part of A truncated towards zero.

IOR (I, J)

IOR is an elemental function, which returns the logical inclusive OR of all bits in I and corresponding bits in J. The arguments I and J must be of type integer with the same kind type parameter. The result is of type integer and has the same kind type parameter as the arguments. Integers which are sequences of bits are interpreted according to the model for bit manipulation. The following truth table shows the effect of the function:

I	1	1	0	0
J	1	0	1	0
IOR(I, J)	1	1	1	0

14

ISHFT (I, SHIFT)

ISHFT is an elemental function, which returns a sequence of bits equal to that of I except that the bits are shifted end-off SHIFT positions. The argument I must be of type integer. The argument SHIFT must be of type integer with |SHIFT| < BIT_SIZE(I). The result is of type integer and has the same kind type parameter as I. If SHIFT > 0, the sequence of bits is shifted to the left. If SHIFT < 0, the sequence of bits is shifted to the right. Bits shifted out to the left or to the right are lost and zeros are inserted into the gaps created. The integers which are sequences of bits are interpreted according to the model for bit manipulation.

ISHFTC (I, SHIFT [, SIZE])

ISHFTC is an elemental function, which returns a sequence of bits equal to that of I except that the SIZE rightmost bits or all bits are shifted circularly SHIFT positions. The argument I must be of type integer. The argument SHIFT must be of type integer with |SHIFT| < BIT_SIZE(I). The optional argument SIZE must be of type integer with $0 <$ SIZE $\leq$ BIT_SIZE(I). If SIZE is absent, the default SIZE = BIT_SIZE(I) is used. The result is of type integer and has the same kind type parameter as I. If SHIFT > 0, the SIZE rightmost bits in I are shifted circularly to the left. If SHIFT < 0, the SIZE rightmost bits in I are shifted circularly to the right. The integers which are sequences of bits are interpreted according to the model for bit manipulation.

KIND (X)

KIND is an inquiry function, which returns the kind type parameter of the argument. The argument X may be scalar or array valued and of any intrinsic type except character. The result is scalar and of type default integer.

LBOUND (ARRAY [, DIM])

LBOUND is an inquiry function, which returns all lower bounds of array ARRAY or which returns the lower bound in dimension DIM. The argument ARRAY must be an array of any type, no scalar. The optional argument DIM must be scalar and of type integer. The result is of type default integer.

If DIM is present, the result is scalar. If ARRAY is an array section or another array expression other than a whole array and a structure component, LBOUND(ARRAY, DIM) returns the value 1. For a whole array or an array structure component, LBOUND(ARRAY, DIM) returns the value of the lower bound for dimension DIM of ARRAY if dimension DIM of ARRAY does not have extent zero; otherwise the result has the value 1.

If DIM is absent, the result is a 1-dimensional array with a size which is equal to the rank of ARRAY. And the values of the array elements of the result are the lower bounds of ARRAY.

LEN (STRING)

LEN is an inquiry function, which returns the character length of the argument. The argument STRING must be of type character. The result is of type default integer.

LEN_TRIM (STRING)

LEN_TRIM is an elemental function, which returns the character length of the argument without its trailing blanks. The argument STRING must be of type character. The result is of type default integer.

LOG (X)

LOG is an elemental function, which returns the natural logarithm (base e). The argument X must be of type real or complex. If X is of type real, X must be positive. If X is of type complex, X must not be zero. The result is of the same type and kind type parameter as X. A result of type complex is the principal value with the imaginary part ω in the range $-\pi < \omega \le \pi$.

LOGICAL (L [, KIND])

LOGICAL is an elemental function, which performs a type (parameter) conversion. The argument L must be of type logical. If KIND is present, the result is of type logical and has the kind type parameter KIND; otherwise the result is of type default logical. The result has the value L.

LOG10 (X)

LOG10 is an elemental function, which returns the common logarithm (base 10). The argument X must be of type real with $X > 0$. The result is of type real with the same kind type parameter as X.

MATMUL (MATRIX_A, MATRIX_B)

MATMUL is a transformational function, which performs matrix multiplication of two numeric or two logical 1-dimensional or 2-dimensional arrays. The argument MATRIX_A must be an array of type numeric or logical. If MATRIX_A is of numeric type, MATRIX_B must be an array of numeric type; and if MATRIX_A is of type logical, MATRIX_B must be an array of type logical. The numeric type of MATRIX_A may be different from the numeric type of MATRIX_B. If MATRIX_A is 1-dimensional, MATRIX_B must be 2-dimensional; and if MATRIX_B is 1-dimensional, MATRIX_A must be 2-dimensional. The extent of the first (or single) dimension of MATRIX_B must be equal to the extent of the last (or single) dimension of MATRIX_A. If the arguments are of numeric type, the type and kind type parameter of the result are those of the expression (MATRIX_A * MATRIX_B). If the arguments are of type logical, the result also is of type logical, and the kind type parameter is that of (MATRIX_A .AND. MATRIX_B). The result is an array. The shape of the result depends on the shapes of the arguments, as follows:

14

If MATRIX_A has the shape (n, m) and MATRIX_B the shape (m, k), the result has the shape (n, k). If the arguments are of numeric type, the array element (i, j) of the result has the value SUM(MATRIX_A$(i, :)$ * MATRIX_B$(:, j)$). And if both arguments are of type logical, the array element (i, j) of the result has the value ANY(MATRIX_A$(i, :)$ .AND. MATRIX_B$(:, j)$).

If MATRIX_A has the shape (m) and MATRIX_B has the shape (m, k), the result has the shape (k). If the arguments are of numeric type, the array element (j) of the result has the value SUM(MATRIX_A$(:)$ * MATRIX_B$(:, j)$). And if both arguments are of type logical, the array element (j) of the result has the value ANY(MATRIX_A$(:)$ .AND. MATRIX_B$(:, j)$).

If MATRIX_A has the shape (n, m) and MATRIX_B has the shape (m), the result has the shape (n). If the arguments are of numeric type, the array element (i) of the result has the value SUM(MATRIX_A$(i, :)$ * MATRIX_B$(:)$). And if both arguments are of type logical, the array element (i) of the result has the value ANY(MATRIX_A$(i, :)$ .AND. MATRIX_B$(:)$).

MAX (A1, A2 [, A3, ...])

MAX is an elemental function, which returns the maximum of two or more given values. The arguments must be of type integer or real; their types and kind type parameters must agree. The result is of the same type and kind type parameter as the arguments.

MAXEXPONENT (X)

MAXEXPONENT is an inquiry function, which returns the maximum exponent in the model that includes the argument X. The argument X must be an array or a scalar of type real. The result is scalar and of type default integer.

MAXLOC (ARRAY [, MASK])

MAXLOC is a transformational function, which determines the first array element containing the maximum value of all elements in array ARRAY or of a given subset of elements in array ARRAY. The argument ARRAY must be an array of type integer or real. The optional argument MASK must be of type logical and must have the same shape as ARRAY. The result is of type default integer.

The result of MAXLOC(ARRAY) is a 1-dimensional array with a size which is equal to the rank of ARRAY and whose element values are the values of the subscripts of an element of ARRAY containing the maximum value of all elements of ARRAY. The result of MAXLOC(ARRAY, MASK=MASK) determines the position of the maximum value of those array elements in ARRAY which correspond to *true* elements in MASK.

Note that in either case not the declared array bounds of ARRAY are used, but value 1 is used as the lower bound in all dimensions of ARRAY.

MAXVAL (ARRAY [, DIM] [, MASK])

MAXVAL is a transformational function, which returns the maximum value of all elements in array ARRAY or of a subset of elements in ARRAY. The argument ARRAY must be an array of type integer or real. The optional argument DIM must be scalar and of type integer. The optional argument MASK must be of type logical and must have the same shape as ARRAY. DIM and MASK may be specified in any order. The result is of the same type and kind type parameter as ARRAY.

If DIM is absent or if ARRAY is a 1-dimensional array, the result is scalar. The value of MAXVAL(ARRAY) is equal to the maximum value of all array elements in ARRAY. The value of MAXVAL(ARRAY, MASK=MASK) is equal to the maximum value of those array elements in ARRAY which correspond to *true* elements in MASK.

If DIM is present, the result is an array of the rank of ARRAY reduced by one and with the shape of ARRAY without the dimension DIM. The function is applied to all 1-dimensional array sections which span right through dimension DIM.

MERGE (TSOURCE, FSOURCE, MASK)

MERGE is an elemental function, which returns a result composed of two given values. The argument TSOURCE may be of any type. FSOURCE must be of the same type, (if applicable) kind type parameter, and (if applicable) character length as TSOURCE. The argument MASK must be of type logical. The result is of the same type, (if applicable) kind type parameter, and (if applicable) character length as TSOURCE. The result is equal to TSOURCE where MASK is *true*; otherwise the result is equal to FSOURCE.

MIN (A1, A2 [, A3, ...])

MIN is an elemental function, which returns the minimum of two or more given values. The arguments must be of type integer or real; their types and kind type parameters must agree. The result is of the same type and kind type parameter as the arguments.

MINEXPONENT (X)

MINEXPONENT is an inquiry function, which returns the minimum exponent in the model that includes the argument X. The argument X must be an array or a scalar of type real. The result is scalar and of type default integer.

MINLOC (ARRAY [, MASK])

MINLOC is a transformational function, which determines the first array element containing the minimum value of all elements in array ARRAY or of a given subset of elements in array ARRAY. The argument ARRAY must be an array of type integer or real. The optional argument MASK must be of type logical and must have the same shape as ARRAY. The result is of type default integer.

The result of MINLOC(ARRAY) is a 1-dimensional array with a size which is equal to the rank of ARRAY and whose element values are the values of the subscripts of an element of ARRAY containing the minimum value of all elements of ARRAY. The result of MINLOC(ARRAY, MASK=MASK) determines the position of the minimum value of those array elements in ARRAY which correspond to *true* elements in MASK.

Note that in either case not the declared array bounds of ARRAY are used, but value 1 is used as the lower bound in all dimensions of ARRAY.

MINVAL (ARRAY [, DIM] [, MASK])

MINVAL is a transformational function, which returns the minimum value of all elements in array ARRAY or of a subset of elements in ARRAY. The argument ARRAY must be an array of type integer or real. The optional argument DIM must be scalar and of type integer. The optional argument MASK must be of type logical and must have the same shape as ARRAY. DIM and MASK may be specified in any order. The result is of the same type and kind type parameter as ARRAY.

If DIM is absent or if ARRAY is an 1-dimensional array, the result is scalar. The value of MAXVAL(ARRAY) is equal to the minimum value of all array elements in ARRAY. The value of MAXVAL(ARRAY, MASK=MASK) is equal to the minimum value of those array elements in ARRAY which correspond to *true* elements in MASK.

If DIM is present, the result is an array of the rank of ARRAY reduced by one and with the shape of ARRAY without the dimension DIM. The function is applied to all 1-dimensional array sections which span right through dimension DIM.

MODULO (A, P)

MODULO is an elemental function, which returns A modulo P. The argument A must be of type integer or real. The types and kind type parameters of the arguments must agree. The result is of the same type and kind type parameter as A. If the arguments are of type integer, the result has the value $(A - FLOOR(REAL(A)/REAL(P)) * P)$. And if the arguments are of type real, the result has the value $(A - FLOOR(A/P) * P)$. If $P=0$, the result is processor-dependent.

MVBITS (FROM, FROMPOS, LEN, TO, TOPOS)

MVBITS is an elemental subroutine, which copies the sequence of bits in FROM that begins at position FROMPOS and has the length LEN to TO beginning at position TOPOS. The arguments FROM, FROMPOS, LEN, and TOPOS have the INTENT(IN) attribute and must be of type integer. FROMPOS must be ≥ 0 with FROMPOS + LEN $\leq$ BIT_SIZE(FROM). LEN must be ≥ 0. TOPOS must be ≥ 0 with TOPOS + LEN $\leq$ BIT_SIZE(TO). The argument TO has the INTENT(INOUT) attribute and must be a variable of type integer and must have the same kind type parameter as FROM. Integers which are sequences of bits are interpreted according to the model for bit manipulation.

NEAREST (X, S)

NEAREST is an elemental function, which returns the nearest different representable machine number in the direction given by the sign of argument S. The argument X must be of type real. The argument S must be of type real; its value must not be zero. The result is of type real with the same kind type parameter as X. If S is positive, the result is the next representable number which is greater than X. And if S is negative, the result is the next representable number which is smaller than X.

NINT (A [, KIND])

NINT is an elemental function, which returns the nearest integer to the argument A. The argument A must be of type real. The result is of type integer. If KIND is present, the result has the kind type parameter KIND; otherwise it is of type default integer.

NOT (I)

NOT is an elemental function, which returns the logical complement of all bits in the argument I. The argument I must be of type integer. The result is of type integer and has the same kind type parameter as I. Integers which are sequences of bits are interpreted according to the model for bit manipulation. The following truth table shows the effect of the function:

I	0	1
NOT(I)	1	0

PACK (ARRAY, MASK [, VECTOR])

PACK is a transformational function, which packs into a 1-dimensional array those elements in the given array ARRAY that are selected by the conformable logical mask MASK. The argument ARRAY may be an array of any type. The argument MASK must be of type logical; it may be scalar or an array, but it must be conformable with ARRAY. The optional argument VECTOR must be a 1-dimensional array of the same type, (if applicable) kind type parameter, and (if applicable) character length as ARRAY. Its size must be greater than or equal to the number of *true* elements in MASK. If MASK is scalar and has the value *true*, VECTOR must have at least the same size as ARRAY.

The result is of the same type, (if applicable) kind type parameter, and (if applicable) character length as ARRAY. If VECTOR is present, the 1-dimensional result has the same size as VECTOR. If VECTOR is absent and MASK is an array, the size of the result is equal to the number of *true* elements in MASK. If VECTOR is absent and MASK is scalar, the size of the result is equal to the size of ARRAY.

The i-th element of the result is equal to that element in ARRAY which corresponds to the i-th *true* element in MASK. The array elements are counted in array element order.

If VECTOR is present and if its size is greater than the number of *true* elements in MASK, the remaining elements of the result have the same value as the remaining elements in VECTOR.

PRECISION (X)

PRECISION is an inquiry function, which returns the decimal precision in the model that includes the argument X. The argument X must be of type real or complex. The result is of type default integer.

PRESENT (A)

PRESENT is an inquiry function, which determines whether argument A, which must be the name of an optional dummy argument that is accessible in the subprogram containing the reference to the PRESENT function, is present in that subprogram. Argument A may be scalar or array valued, of any type, a pointer, or a dummy subprogram. Dummy argument A has no INTENT attribute. The result is of type default logical. The result has the value *true*, if A is present; otherwise the result has the value *false*.

PRODUCT (ARRAY [, DIM] [, MASK])

PRODUCT is a transformational function, which returns the product of all elements in array ARRAY or of a subset of elements in ARRAY. The argument

ARRAY must be an array of numeric type. The optional argument DIM must be scalar and of type integer. The optional argument MASK must be of type logical and must have the same shape as ARRAY. The result is of the same type and kind type parameter as ARRAY.

If MASK and DIM are absent, the result is scalar and its value is the product of all array elements in ARRAY. If MASK is absent and DIM is present, the result is an array (or a scalar if ARRAY is 1-dimensional) of the rank of ARRAY reduced by one and with the shape of ARRAY without the dimension DIM. The function is applied to all 1-dimensional array sections which span right through dimension DIM.

If MASK is present, the function is applied at most to those elements in ARRAY which correspond to *true* elements in MASK.

RADIX (X)

RADIX is an inquiry function, which returns the base of the model that includes the argument X. The argument X must be of type integer or real and may be scalar or an array. The result is of type default integer. The result has the value r if X is of type integer and the value b if X is of type real (see appendix B).

RANDOM_NUMBER (HARVEST)

RANDOM_NUMBER is subroutine, which returns through the argument HARVEST one pseudorandom number or an array of pseudorandom numbers from the uniform distribution over the range $0 \le x < 1$. The argument HARVEST must be a variable of type real, it may be scalar or an array, and it has the INTENT(OUT) attribute.

RANDOM_SEED ([SIZE] [, PUT] [, GET])

14

RANDOM_SEED is a subroutine, which initializes and restarts the pseudorandom number generator and allows inquiries. The pseudorandom numbers may be generated by the invocation of the subroutine RANDOM_NUMBER. The generation of pseudorandom numbers may be suspended at any point of this sequence of pseudorandom numbers and may be resumed later exactly at this point. Either one argument may be present or no argument may be present. If no argument is present, the F processor initializes the seed with a processor-dependent value. SIZE is an optional argument with INTENT(OUT) attribute; it must be a scalar variable of type default integer, which returns the size N of the seed array. PUT is an optional argument with INTENT(IN) attribute; it must be a 1-dimensional array of type default integer and size $\ge N$, which is used to reset the seed. GET is an optional argument with INTENT(OUT) attribute; it must be a 1-dimensional array variable of type default integer and size $\geq N$, which returns the current value of the seed. The value of GET may be later used to restart the seed.

Initialization of the pseuderandom number generator by a particular PUT argument is performed in a processor-dependent manner. The value returned by GET need not be the same as the value of PUT in an immediately preceding invocation of RANDOM_SEED. Though these values differ, when used as PUT argument in RANDOM_SEED the pseudorandom number sequences generated by RANDOM_NUMBER are equal.

RANGE (X)

RANGE is an inquiry function, which returns the decimal exponent range in the model that includes the argument X. The argument X must be of type numeric; it may be scalar or an array. X need not be defined. The result is scalar and of type default integer. If X is of type integer, the result has the value INT(LOG10(HUGE(X))). If X is of type real or complex, the result has the value INT(MIN(LOG10($huge$)), –LOG10($tiny$))), where $huge$ is the largest positive number and $tiny$ the smallest number in the model representing real numbers with the same kind type parameter as X.

REAL (A [, KIND])

REAL is an elemental function, which performs a type conversion. The argument must be of numeric type. If KIND is present, the result is of type real and has the kind type parameter KIND. If KIND is not present and if A is of type integer or real, the result is of type default real and its value is a processor-dependent approximation to A. If KIND is not present and if A is of type complex, the result is of type real and has the same kind type parameter as A and its value is a processor-dependent approximation to the real part of A.

REPEAT (STRING, NCOPIES)

REPEAT is a transformational function, which concatenates several given copies of a string. The argument STRING must be scalar and of type character. The argument NCOPIES must be nonnegative, scalar, and of type integer. The result is scalar and of type character. The result is the string which consists of NCOPIES concatenated copies of STRING.

RESHAPE (SOURCE, SHAPE [, PAD] [, ORDER])

RESHAPE is a transformational function, which returns an array with shape SHAPE, which is of the same type, (if applicable) kind type parameter, and (if applicable) character length as SOURCE. SOURCE must be an array of any type. SHAPE must be a 1-dimensional array of type integer with at most 7 elements. The array elements in SHAPE must all be ≥ 0. If PAD is absent or of size zero, the size of SOURCE must be greater than or equal to the product of the elements in SHAPE (i. e., PRODUCT(SHAPE)). The size of the result is the product of the elements in SHAPE. The optional argument PAD must be an array and of the same type, (if applicable) kind type parameter, and (if applicable) character length as SOURCE. The optional argument ORDER must be a 1-dimensional array of type integer with the same shape as SHAPE. The value of ORDER must be a permutation of $(1, 2, ... , n)$, where n is the size of SHAPE.

If ORDER is absent, the values of the elements of the result are, in array element order, equal to the values of the elements in SOURCE in array element order followed by copies of PAD in array element order. If ORDER is present, the values of the elements of the result correspond in the permuted subscript order (ORDER(1), ORDER(2), ..., ORDER(n)) to the values of the elements in SOURCE in array element order followed by copies of PAD in array element order.

RRSPACING (X)

RRSPACING is an elemental function, which returns the reciprocal of the relative spacing of model numbers that include X near the argument value. The argument X must be of type real. The result is of type real with the same kind type parameter as X. The result has the value $|X \times b^{-e}| \times b^p$ (see appendix B).

SCALE (X, I)

SCALE is an elemental function, which returns the value of the argument X scaled by b^I, where b is the base in the model that includes X. The argument X must be of type real. The result is of type real with the same kind type parameter as X. The result has the value $X \times b^I$ (see appendix B).

SCAN (STRING, SET [, BACK])

SCAN is an elemental function, which returns the position of a character in STRING that is in SET. The arguments STRING and SET must be of type character. The optional argument BACK must be of type logical. The result is of type default integer. The result is 0 if no character in STRING is in SET or if the length of STRING or SET is zero. Otherwise: if BACK is absent or present with the value *false*, the result is the position of the leftmost character in STRING that is in SET; and if BACK is present with the value *true*, the result is the position of the rightmost character in STRING that is in SET.

SELECTED_INT_KIND (R)

SELECTED_INT_KIND is a transformational function, which returns a kind type parameter value of an integer type for the representation of all integers n in the range $-10^R < n < 10^R$. The argument R must be scalar and of type integer. The result is scalar and of type default integer. If the F processor does not support such integer type, the result has the value -1. If the F processor supports more than one integer type for the given range, the result has the value of the kind type parameter with the smallest decimal exponent range. If the F processor supports more than one integer type for the given range with the same smallest decimal exponent range, the result has the value of the smallest value of their kind type parameters.

SELECTED_REAL_KIND ([P] [, R])

SELECTED_REAL_KIND is a transformational function, which returns a kind type parameter value of a real type for the representation of all reals with a decimal precision of at least P digits (as returned by the intrinsic function PRECISION) and a decimal exponent range of at least R (as returned by the intrinsic function RANGE). If the F processor does not support such real type, the result has the value -1 if the minimum precision P is not supported, the value -2 if the minimum exponent range R is not supported, and the value -3 if neither the given P nor the given R are supported. If the F processor supports more than one real type for the given arguments, the result has the value of the kind type parameter with the smallest decimal precision. If the F processor supports more than one real type for the given arguments with the same smallest decimal precision, the result has the value of the smallest value of their kind type parameters.

SET_EXPONENT (X, I)

SET_EXPONENT is an elemental function, which returns the number in the model that includes X whose fractional part is the fractional part of the model representation of argument X and whose exponent part is the argument I. The argument X must be of type real. The argument I must be of type integer. The result is of type real with the same kind type parameter as X. The result has the value $(X \times b^{I-e})$ (see appendix B).

SHAPE (SOURCE)

SHAPE is an inquiry function, which returns the shape of the argument SOURCE. The argument SOURCE may be scalar or an array of any type; it must not be a currently disassociated pointer or a currently deallocated allocatable array. The result is a 1-dimensional array of type default integer,

whose size is equal to the rank of source. The value of the result is the shape of
SOURCE. If SOURCE is scalar, the result has size zero.

SIGN (A, B)

SIGN is an elemental function, which returns the absolute value of the argument
A times the sign of argument B. The arguments A and B must be of type integer
or real; their types and kind type parameters must agree. The result is of the
same type and kind type parameter as A.

SIN (X)

SIN is an elemental function, which returns the sine. The argument X must be
of type real or complex and must be specified in radians. The result is of the
same type and kind type parameter as X.

SINH (X)

SINH is an elemental function, which returns the hyperbolic sine. The argument
X must be of type real. The result is of type real with the same kind type
parameter as X.

SIZE (ARRAY [, DIM])

SIZE is an inquiry function, which returns the size of the array argument
ARRAY or the size of ARRAY along dimension DIM. The argument ARRAY
may be an array of any type, but no scalar. The optional argument DIM must
be scalar and of type integer. The result is scalar and of type default integer. If
DIM is absent, the result is the size of ARRAY. If DIM is present, the result is
the extent of dimension DIM of ARRAY.

14

SPACING (X)

SPACING is an elemental function, which returns the absolute spacing of the
model numbers that include X near the argument value. The argument must
be of type real. The result is of type real with the same kind type parameter as
X. If X is not zero, the result has the value b^{e-p}, provided this result is within
range (see appendix B). Otherwise, the result is the same as that of TINY(X).

SPREAD (SOURCE, DIM, NCOPIES)

SPREAD is a transformational function, which returns an array with the rank
of SOURCE increased by one that consists along its given dimension DIM of
NCOPIES copies of SOURCE. The argument SOURCE may be scalar or an
array of any type. The arguments DIM and NCOPIES must be scalar and of

type integer. The result is of the same type, (if applicable) kind type parameter, and (if applicable) character length as SOURCE.

If SOURCE is scalar, the result is a 1-dimensional array with MAX(NCOPIES, 0) elements that have the same value as SOURCE. If SOURCE is an array with the shape $(d_1, d_2, \ldots, d_n)$, the shape of the result is $(d_1, d_2, \ldots, d_{DIM-1}, \mathrm{MAX(NCOPIES, 0)}, d_{DIM}, \ldots, d_n)$. And the array element of the result with the subscript $(r_1, r_2, \ldots, r_{n+1})$ has the value $\mathrm{SOURCE}(r_1, r_2, \ldots, r_{DIM-1}, r_{DIM+1}, \ldots, r_{n+1})$.

SQRT (X)

SQRT is an elemental function, which returns the square root $\sqrt{X}$. The argument X must be of type real or complex. If X is of type real, it must not be negative. The result is of the same type and kind type parameter as X. If X is of type complex, the result is the principal value with the real part greater than or equal to zero. When the real part of the result is zero, the imaginary part of the result is not negative.

SUM (ARRAY [, DIM] [, MASK])

SUM is a transformational function, which returns the sum of all elements in array ARRAY or of a subset of elements in ARRAY. The argument ARRAY must be an array of numeric type. The optional argument DIM must be scalar and of type integer. The optional argument MASK must be of type logical and must have the same shape as ARRAY. The result is of the same type and kind type parameter as ARRAY.

If MASK and DIM are absent, the result is scalar and its value is the sum of all array elements in ARRAY. If MASK is absent and DIM is present, the result is an array (or a scalar if ARRAY is 1-dimensional) of the rank of ARRAY reduced by one and with the shape of ARRAY without the dimension DIM. The function is applied to all 1-dimensional array sections which span right through dimension DIM.

If MASK is present, the function is applied at most to those elements in ARRAY which correspond to *true* elements of MASK.

SYSTEM_CLOCK([COUNT][,COUNT_RATE][,COUNT_MAX])

SYSTEM_CLOCK is an intrinsic subroutine, which returns through its output arguments information from the real-time clock. All arguments are optional arguments with INTENT(OUT) attribute and must be scalar variables of type default integer.

COUNT is set to the processor-dependent value of the system clock. This system clock is incremented by one at each clock count until COUNT_MAX is reached and is set to zero on the next count. If the F processor does not support a system clock, COUNT is set to −HUGE(0).

COUNT_RATE is set to the processor-dependent approximation to the number of processor clock counts per second. If the F processor does not support a system clock, COUNT_RATE is set to zero.

COUNT_MAX is set to the maximum value that COUNT may have. If the F processor does not support a system clock, COUNT_MAX is set to zero.

TAN (X)

TAN is an elemental function, which returns the tangent. The argument X must be of type real and must be specified in radians. The result is of type real with the same kind type parameter as X.

TANH (X)

TANH is an elemental function, which returns the hyperbolic tangent. The argument X must be of type real. The result is of type real with the same kind type parameter as X.

TINY (X)

TINY is an inquiry function, which returns the smallest number in the model that includes X. The argument X may be a scalar or an array and must be of type real. The result is scalar and is of type real with the same kind type parameter as X. The result has the value $b^{e_{min}-1}$ (see appendix B).

TRANSPOSE (MATRIX)

14

TRANSPOSE is a transformational function, which performs a matrix transpose for a given 2-dimensional array. The argument MATRIX may be a 2-dimensional array of any type. The result is of the same type, (if applicable) kind type parameter and (if applicable) character length as MATRIX. The result is a 2-dimensional array with the shape (n, m), where (m, n) is the shape of MATRIX. The array element (i, j) of the result has the value MATRIX(j, i) for $i = 1, 2, ..., n$ and $j = 1, 2, ..., m$.

TRIM (STRING)

TRIM is a transformational function, which returns the string STRING without its trailing blanks. The argument STRING must be scalar and of type character. The result is of type character. The length of the result is equal to the length of STRING reduced by the number of trailing blanks in STRING.

UBOUND (ARRAY [, DIM])

UBOUND is an inquiry function, which returns all upper bounds of array ARRAY or which returns the upper bound in dimension DIM. The argument ARRAY must be an array of any type, no scalar. The optional argument DIM must be scalar and of type integer. The result is of type default integer.

If DIM is present, the result is scalar. If ARRAY is an array section or another array expression which is no whole array and no structure component, the result is equal to the extent of the dimension DIM. If the extent of the dimension DIM of ARRAY is not zero, the result is the value of the upper bound in dimension DIM. And if the extent of the dimension DIM of ARRAY is zero, the result is 0.

If DIM is absent, the result is a 1-dimensional array with a size which is equal to the rank of ARRAY. And the values of the array elements of the result are the upper bounds of ARRAY.

UNPACK (VECTOR, MASK, FIELD)

UNPACK is a transformational function, which unpacks a 1-dimensional array into an array of a given type and a given shape. The argument VECTOR may be a 1-dimensional array of any type. Its size must be greater than or equal to the number of *true* elements in MASK. The argument MASK must be an array of type logical. The argument FIELD must be of the same type, (if applicable) kind type parameter, and (if applicable) character length as VECTOR. FIELD may be scalar or an array, but it must be conformable with MASK.

The result is an array and is of the same type, (if applicable) kind type parameter, and (if applicable) character length as VECTOR. It has the same shape as MASK. The array element of the result which corresponds to the ith *true* element in MASK has the value VECTOR(i) for $i = 1, 2, ..., t$, where t is the number of *true* elements in MASK. The array elements are counted in array element order.

All other elements of the result are equal to the value of FIELD if FIELD is scalar, or equal to the corresponding element in FIELD if FIELD is an array.

VERIFY (STRING, SET [, BACK])

VERIFY is an elemental function, which returns the value zero if each character in STRING is in SET, or which returns the position of a character of STRING which is *not* in SET. If the optional argument BACK is absent or if it is present with the value *false*, the result is the position of the leftmost such character which is not in SET. And if BACK is present with the value *true*, the result is the position of the rightmost such character which is not in SET.

A ASCII CHARACTER SET AND COLLATING SEQUENCE

Bits b4 b3 b2 b1	Control Characters		Symbols, Digits		Upper-case Letters		Lower-case Letters	
0 0 0 0	NUL (0, 0, 0)	DLE (16, 10, 20)	SP (32, 20, 40)	0 (48, 30, 60)	@ (64, 40, 100)	P (80, 50, 120)	` (96, 60, 140)	p (112, 70, 160)
0 0 0 1	SOH (1, 1, 1)	DC1 (17, 11, 21)	! (33, 21, 41)	1 (49, 31, 61)	A (65, 41, 101)	Q (81, 51, 121)	a (97, 61, 141)	q (113, 71, 161)
0 0 1 0	STX (2, 2, 2)	DC2 (18, 12, 22)	" (34, 22, 42)	2 (50, 32, 62)	B (66, 42, 102)	R (82, 52, 122)	b (98, 62, 142)	r (114, 72, 162)
0 0 1 1	ETX (3, 3, 3)	DC3 (19, 13, 23)	# (35, 23, 43)	3 (51, 33, 63)	C (67, 43, 103)	S (83, 53, 123)	c (99, 63, 143)	s (115, 73, 163)
0 1 0 0	EOT (4, 4, 4)	DC4 (20, 14, 24)	$ (36, 24, 44)	4 (52, 34, 64)	D (68, 44, 104)	T (84, 54, 124)	d (100, 64, 144)	t (116, 74, 164)
0 1 0 1	ENQ (5, 5, 5)	NAK (21, 15, 25)	% (37, 25, 45)	5 (53, 35, 65)	E (69, 45, 105)	U (85, 55, 125)	e (101, 65, 145)	u (117, 75, 165)
0 1 1 0	ACK (6, 6, 6)	SYN (22, 16, 26)	& (38, 26, 46)	6 (54, 36, 66)	F (70, 46, 106)	V (86, 56, 126)	f (102, 66, 146)	v (118, 76, 166)
0 1 1 1	BEL (7, 7, 7)	ETB (23, 17, 27)	' (39, 27, 47)	7 (55, 37, 67)	G (71, 47, 107)	W (87, 57, 127)	g (103, 67, 147)	w (119, 77, 167)
1 0 0 0	BS (8, 8, 10)	CAN (24, 18, 30)	((40, 28, 50)	8 (56, 38, 70)	H (72, 48, 110)	X (88, 58, 130)	h (104, 68, 150)	x (120, 78, 170)
1 0 0 1	HT (9, 9, 11)	EM (25, 19, 31)	) (41, 29, 51)	9 (57, 39, 71)	I (73, 49, 111)	Y (89, 59, 131)	i (105, 69, 151)	y (121, 79, 171)
1 0 1 0	LF (10, A, 12)	SUB (26, 1A, 32)	* (42, 2A, 52)	: (58, 3A, 72)	J (74, 4A, 112)	Z (90, 5A, 132)	j (106, 6A, 152)	z (122, 7A, 172)
1 0 1 1	VT (11, B, 13)	ESC (27, 1B, 33)	+ (43, 2B, 53)	; (59, 3B, 73)	K (75, 4B, 113)	[(91, 5B, 133)	k (107, 6B, 153)	{ (123, 7B, 173)
1 1 0 0	FF (12, C, 14)	FS (28, 1C, 34)	, (44, 2C, 54)	< (60, 3C, 74)	L (76, 4C, 114)	\ (92, 5C, 134)	l (108, 6C, 154)	| (124, 7C, 174)
1 1 0 1	CR (13, D, 15)	GS (29, 1D, 35)	- (45, 2D, 55)	= (61, 3D, 75)	M (77, 4D, 115)	] (93, 5D, 135)	m (109, 6D, 155)	} (125, 7D, 175)
1 1 1 0	SO (14, E, 16)	RS (30, 1E, 36)	. (46, 2E, 56)	> (62, 3E, 76)	N (78, 4E, 116)	^ (94, 5E, 136)	n (110, 6E, 156)	~ (126, 7E, 176)
1 1 1 1	SI (15, F, 17)	US (31, 1F, 37)	/ (47, 2F, 57)	? (63, 3F, 77)	O (79, 4F, 117)	_ (95, 5F, 137)	o (111, 6F, 157)	DEL (127, 7F, 177)

b7 b6 b5 column bits:
- Control Characters: 0 0 0 / 0 0 1
- Symbols, Digits: 0 1 0 / 0 1 1
- Upper-case Letters: 1 0 0 / 1 0 1
- Lower-case Letters: 1 1 0 / 1 1 1

A

Legend:

dec	
char	
hex	oct

The ASCII character set is a standard character set. The F language refers to ISO/IEC 646:1991. With regard to letters, digits, and special characters of the F character set, ISO/IEC 646:1991 and ANSI X3.4-1986, which is the US national version of the ISO/IEC standard, are identical.

An ASCII character is encoded by 7 bits. Therefore, a standard ASCII character set contains a total number of 128 characters. The first 32 characters (hex 00 to hex 1F) of the ASCII character set are *control characters* and *graphic characters*. Nearly all other characters (hex 20 to hex 7E) are *printable* characters.

The ASCII character set has a collating sequence. The first character has position 0, the second has position 1, and so on.

The ICHAR intrinsic function returns for a given character the position of this character in the ASCII collating sequence; ICHAR("X") returns the integer value 88. And the CHAR intrinsic function returns for a given nonnegative integer value the character that corresponds to this given value in the ASCII collating sequence; CHAR(88) returns the character "X".

B   MODELS FOR NUMBERS

The bit manipulation intrinsic subprograms, the numeric manipulation intrinsic functions, and the inquiry intrinsic functions are defined in terms of models for the representation of each kind of integer or real data implemented by the processor. These models are used to describe the characteristics and the behaviour of the corresponding number sets of a particular type. A model has parameters which are chosen by the F processor such that the model best fits the hardware.

Note that these models do not dictate to the F processor how it has to implement numbers and how they should behave, but the internal representation and the behaviour of the actually implemented numbers are described in terms of these (abstract) models.

B.1   Models for Integers

The models for integer data i are defined by:

$$i = s \times \sum_{k=1}^{q} w_k \times r^{k-1}$$

with

	Value	Description
s	+1 or -1	sign
r	integer > 1	base
q	integer > 0	max. number of digits
w_k	each an integer with $0 \le w_k < r$	digit

The parameters r and q determine the set of values of the integer model numbers.
Example: For model $i = s \times \sum_{k=1}^{31} w_k \times 2^{k-1}$ is $q = 31$ and $r = 2$.

B

B.2   Models for Reals

The models for real data x are defined by:

$$x = \begin{cases} 0 & (= b^0 \times \sum_{k=1}^{p} 0 \times b^{-k}) \\ s \times b^e \times \sum_{k=1}^{p} f_k \times b^{-k} \end{cases}$$

with

	Value	Description
s	+1 or −1	sign
b	integer > 1	base
p	integer > 1	max. number of digits
f_k	each an integer with $0 \leq f_k < b$, but $f_1 > 0$	digit
e	integer with $e_{min} \leq e \leq e_{max}$	exponent

The parameters b, p, e_{min}, and e_{max} determine the set of values of the (real) floating point model numbers.

Example: For model $x = \begin{cases} 0 \\ s \times 2^e \times (1/2 + \sum_{k=2}^{24} f_k \times 2^{-k}), & -126 \leq e \leq 127 \end{cases}$
is $b = 2$, $p = 24$, $e_{min} = -126$, and $e_{max} = 127$.

B.3 Models for Bit Manipulation

Bit manipulations are performed on integer data. Therefore, a **bit** is defined as a binary digit w at position k of a nonnegative integer datum. The models for these integers are defined by:

$$j = \sum_{k=0}^{s-1} w_k \times 2^k$$

with

	Value	Description
s	integer > 0	maximal number of digits
w_k	each either 0 or 1	digit, bit

The parameter s determines the set of values.

Example: For model $j = \sum_{k=0}^{31} w_k \times 2^k$ is $s = 32$. It defines a model for integer data with 32 bits.

The models for bit manipulation define that the integer datum is a sequence of s bits, which are numbered from the right to the left beginning with 0 for the rightmost bit and ending with $(s-1)$ for the leftmost bit. These models apply only to the intrinsic subprograms for bit manipulation and bit inquiry. For all other purposes, the "normal" models for integer data must be used.

C   PROGRAM EXAMPLE

```
! Copyright (c) 1994 Unicomp, Inc.
!
! Developed at Unicomp, Inc.
!
! Permission to use, copy, modify, and distribute this
! software is freely granted, provided that this notice
! is preserved.

module tree_sort_module

public :: insert, print_tree

type, public :: node
   integer :: value
   type (node), pointer :: left, right
end type node

integer, public :: number

contains

   recursive subroutine insert (t)

      type (node), pointer :: t   ! A tree

      ! If (sub)tree is empty, put number at root
      if (.not. associated (t)) then
         allocate (t)
         t % value = number
         nullify (t % left)
         nullify (t % right)
      ! Otherwise, insert into correct subtree
      else if (number < t % value) then
         call insert (t % left)
      else
         call insert (t % right)
      end if

   end subroutine insert
```

```
recursive subroutine print_tree (t)
! Print tree in infix order

   type (node), pointer :: t   ! A tree

   if (associated (t)) then
      call print_tree (t % left)
      print *, t % value
      call print_tree (t % right)
   end if

end subroutine print_tree

end module tree_sort_module

program tree_sort
! Sorts a file of integers by building a
! tree, sorted in infix order.
! This sort has expected behavior n log n,
! but worst case (input is sorted) n ** 2.

   use tree_sort_module

   type (node), pointer :: t   ! A tree
   integer :: ios

   nullify (t)   ! Start with empty tree
   do
      read (unit=*, fmt=*, iostat = ios) number
      if (ios < 0) then
         exit
      end if
      call insert (t) ! Put next number in tree
   end do
   ! Print nodes of tree in infix order
   call print_tree (t)

end program tree_sort
```

You will find this and several other examples at
http://www.imagine1.com/imagine1/example_code.html

D   F versus FORTRAN 90

The following list gives a rough overview of those Fortran language elements and concepts not being defined/supported in F:

- alternate RETURN
- ampersand as first nonblank character in a line
- apostroph-delimited character literal constant
- arithmetic IF statement
- array (dimension) specification after the ::
- ASSIGN statement, assigned GOTO, and assigned format specifier
- assumed length function
- assumed size array
- asterisk dummy argument
- attribute specification statements: ALLOCATABLE, DATA, DIMENSION, INTENT, OPTIONAL, PARAMETER, POINTER, TARGET, SAVE
- BLOCK DATA program unit
- BOZ literal constants
- branching to END IF from outside the IF construct
- character length specification after the ::
- CALL statement in function
- character set other than the ASCII character set
- character string edit descriptors
- COMMON
- computed GOTO
- CONTINUE statement
- default length for character object
- derived-type definition outside the specification part of a module
- DO statement with comma after the DO
- DO termination other than END DO
- DOUBLE PRECISION (except via KIND)
- DO variable being a dummy argument, a pointer, a result variable, or accessed by host association or USE association
- DO WHILE loop

- edit descriptors B, BN, BZ, D, E, EN, G, H, O, P, X, and Z
- ENTRY statement
- EQUIVALENCE statement
- fixed (FORTRAN 77) source form
- FORMAT statement
- function with side effect (except STOP, formatted i/o with standard units)
- generic interface block outside the specification part of a module
- GOTO statement
- implicit interface
- IMPLICIT statement, except IMPLICIT NONE in main program or module
- implicit typing
- implied DO input/output list item
- INCLUDE line
- initialization of variable in main program or function
- input/output positioning statements (BACKSPACE, REWIND, ENDFILE) without parentheses
- i/o specifiers BLANK=, DELIM=, END=, ERR=, NML=, and PAD=
- internal subprogram
- INTRINSIC as an attribute in a type declaration statement
- intrinsic functions ACHAR, DBLE, DIM, DPROD, IACHAR, LGE, LGT, LLE, LLT, MOD, and TRANSFER
- intrinsic subprogram name as actual argument
- kind type parameter for character type
- logical case expression and selector
- logical IF statement
- names with final underscore
- NAMELIST input/output
- optional comma in DO statement
- PAUSE statement
- real DO variable
- relational operators in character form: .EQ., .LT., etc.
- sequence association of arguments
- SEQUENCE statement and sequence type

- shared DO termination
- splitting of lexical tokens onto two lines, even if embedded blanks are allowed
- statement function
- statement label
- statement separator
- stop-code in STOP statement
- storage sequence
- storage unit
- type specifier in FUNCTION statement
- USE statement with empty ONLY list
- WHERE statement (but WHERE construct is allowed)

The following list gives a rough overview of those language elements which are required for an F program:

- accessibility attribute must appear in TYPE definition statement
- accessibility attribute must be specified for module entities
- attributes of a data object must appear in its type declaration
- CALL statement must have parentheses
- CHARACTER statement must be written with a length specifier
- CLOSE statement must be executed before reopening a file
- comma as a separator in a format specification
- complex literal constant must be written as a pair of real literal constants with the same kind type parameter (if any)
- DEFAULT selector, if any, must appear as the final selector
- derived type definition must appear in a module
- dummy arguments of a function must have INTENT(IN) unless it is a pointer or a subprogram
- dummy arguments of type character must have assumed length
- dummy nonpointer arrays must be assumed-shape arrays
- dummy subprogram in a function must also be a function
- END statement must be written in its longest form, for instance,
  END PROGRAM *name*, END FUNCTION *name*, etc.

- external subprogram reference is allowed only where an interface block with an interface definition of the external subrogram is available

- F function must have no side effects, except it may have STOP and certain input/output statement for standard units

- F subprogram must be in a module

- FUNCTION statement must have an RESULT clause

- generic interface block must appear in a module

- i/o specifiers must be written with a leading keyword= such as UNIT=

- kind type parameter values must be written as named constants

- lexical tokens must be written using lower case

- main program must have a PROGRAM statement

- module entities must have an explicitly specified PUBLIC or PRIVATE attribute

- modules have a particular form

- named data object must be declared explicitly

- nonpointer dummy arguments (except subprograms) must have an explicitly specified INTENT attribute

- OPEN must have POSITION=, STATUS=, and ACTION= specifiers

- optional arguments, if any, must follow all non-optional arguments

- processor character set is the ASCII character set

- real literal constant must have the n.n form (optionally with a subsequent exponent and/or kind type parameter)

- SAVE attribute must be specified for any initialized variable

- specification statements must have the :: where it is optional in Fortran

- statements in the specification part of a main program, module, and module subprogram must appear in a predefined order

- user-defined name must be different from any reserved word

- user-defined names are case-sensitive

- user-defined subprogram must be embedded in a module

- SUBROUTINE statement must be written with parentheses

E   SYNTAX CHARTS

The syntax of the F language is described here graphically in terms of railroad diagrams. These charts are just another form of the syntax and are designed only for readability.

The railroad diagrams are not totally equivalent with the original BNF syntax of the F language as described in http://www.imagine1.com/imagine1/bnf.html[1]. But normally those language features that cannot be represented by BNF rules also cannot be represented by railroad diagrams; for example: though the ordering of statements can be represented by railroad diagrams, they cannot describe how to split a complete program, program unit, or subprogram into source lines. To increase readability of the charts, upper-case letters are used to represent certain lexical tokens; i. e., the charts do not show the correct use of upper-case and lower-case letters.

Neither the BNF rules nor the railroad diagrams are a complete and accurate description of the F language, and cannot be used as a basis for parsing. Therefore, where a rule/chart is incomplete, it is accompanied by one or more constraints.

The BNF syntax of the F language is derived from the Fortran 90 syntax [6]. That's why the railroad diagrams within this appendix are classified and numbered according to the chapters in the standard document.

E.1   Notation Used in this Syntax

Characters from the F character set written in **this font** mostly form terminal symbols and are interpreted literally. Italicized letters and words (often hyphenated and abbreviated) represent general syntactic classes for which specific syntactic entities must be substituted in actual statements.

Common abbreviations used in syntactic terms are:

stmt	for	statement	*attr*	for	attribute
expr	for	expression	*decl*	for	declaration
op	for	operator	*desc*	for	descriptor
int	for	integer	*spec*	for	specifier
arg	for	argument	*def*	for	definition

The syntactic classes *letter*, *digit*, and *special-character* are defined as in the F character set (see page 1-1).

---

[1]) Or via Imagine1 home page http://www.imagine1.com/imagine1

In order to minimize the number of additional railroad diagrams and convey appropriate constraint information, the following diagrams are assumed. The letters "*xyz*" stand for any legal syntactic class phrase:

xyz-list

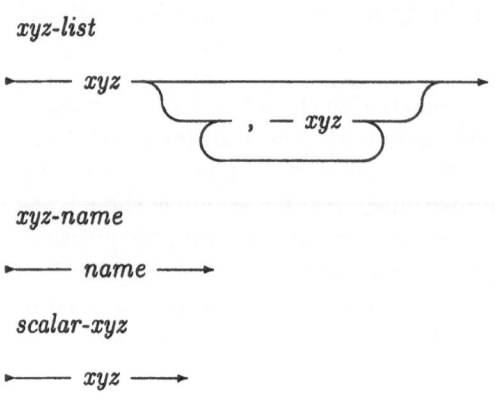

xyz-name

▶────── *name* ──────▶

scalar-xyz

▶────── *xyz* ──────▶

Constraint: *scalar-xyz* shall be scalar.

E.2 F Terms and Concepts

R201 *program*

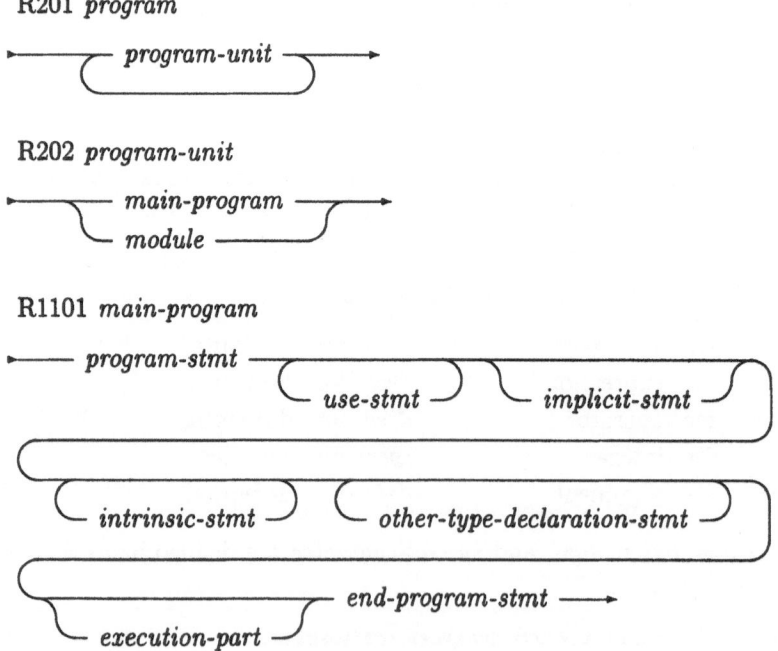

R202 *program-unit*

R1101 *main-program*

R208 *execution-part*

R215 *executable-construct*

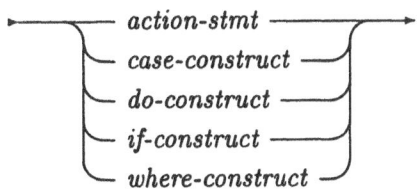

R216 *action-stmt*

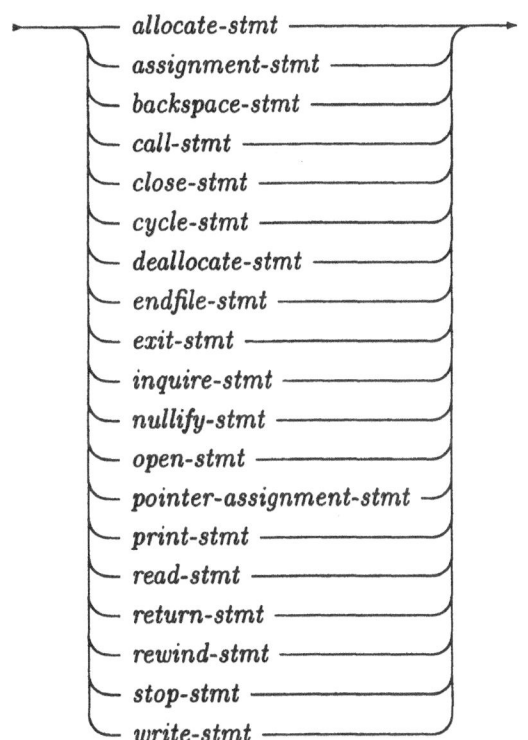

E.3   Characters, Lexical Tokens, and Source Form

R301 *character*

- alphanumeric-character
- special-character

R302 *alphanumeric-character*

- letter
- digit
- underscore

R303 *underscore*

_

R304 *name*

letter
alphanumeric-character $_{max.\ 30}$

R304x *historical-name*

- historical-operator-name
- unsupported-intrinsic-name
- potential-name

R304y *historical-operator-name*

- eq
- ne
- gt
- ge
- lt
- le

R304z *potential-name*

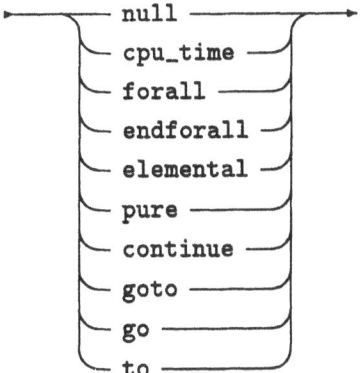

Constraint: No name, regardless of case, shall be the same as an *historical-name*, logical literal constant, logical operator, or intrinsic procedure name.

Constraint: No name, regardless of case, shall be the same as a statement keyword except that they may be the same as a keyword whose only use is as an argument specifier in a *spec-list*, or the names **stat** or **iolength**.

Constraint: The maximum length of a name is 31 characters.

Constraint: The last character of a name shall not be _ .

Constraint: All variables must be declared in type statements or accessed by USE or host association.

Constraint: Entity names, type names, defined operator names, argument keywords for non-intrinsic procedures, and non-intrinsic procedure names may be in mixed upper and lower case, however all references to the names shall use the same case convention.

Constraint: All intrinsic procedure names and argument names for intrinsic procedures shall be lower case.

Constraint: All procedures shall have an explicit interface.

Constraint: All statement keywords shall be in lower case.

Constraint: The logical literal constants .TRUE. and .FALSE. and the logical operators .NOT., .OR., .AND., .EQV., and .NEQV. shall be in lower case.

Constraint: Blank characters shall not appear within any name, keyword, operator, delimiter, or *literal-constant* except that one or more blank characters may appear before or after the *real-part* or *imag-part* of a *complex-literal-constant* and one or more blanks may be used in keywords as follows:

Keyword	Alternate Usage
elseif	else if
enddo	end do
endfile	end file
endfunction	end function
endif	end if
endinterface	end interface
endmodule	end module
endprogram	end program
endselect	end select
endsubroutine	end subroutine
endtype	end type
endwhere	end where
inout	in out
selectcase	select case

Constraint: No name, keyword, delimiter, or operator, shall be split onto more than one line via statement continuation. Keywords shall not be continued at the optional blank.

Constraint: No line shall begin with the **&** character.

R305 *constant*

R306 *literal-constant*

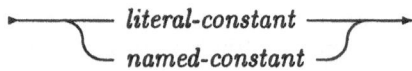

R307 *named-constant*

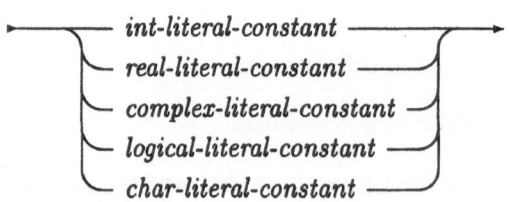

R308 *int-constant*

▸—— *constant* ——▸

Constraint: *int-constant* shall be of type integer.

R309 *char-constant*

▸—— *constant* ——▸

Constraint: *char-constant* shall be of type character.

R310 *intrinsic-operator*

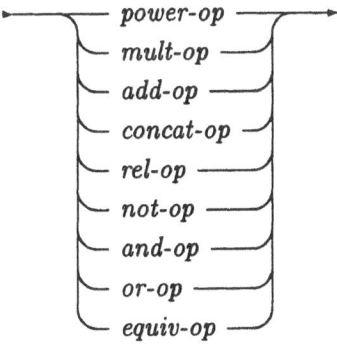

R311 *defined-operator*

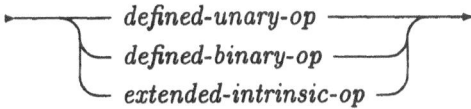

R312 *extended-intrinsic-op*

▸—— *intrinsic-operator* ——▸

Constraint: A *defined-unary-op* and a *defined-binary-op* shall not contain more than 31 letters and shall not be the same as any *intrinsic-operator* or *logical-literal-constant.*

E

E.4 Intrinsic and Derived Data Types

R401 *signed-digit-string*

R402 *digit-string*

R403 *signed-int-literal-constant*

R404 *int-literal-constant*

R405 *kind-param*

R406 *sign*

Constraint: The value of *kind-param* shall be nonnegative.

Constraint: The value of *kind-param* shall specify a representation method that exists on the processor.

R412 *signed-real-literal-constant*

R413 *real-literal-constant*

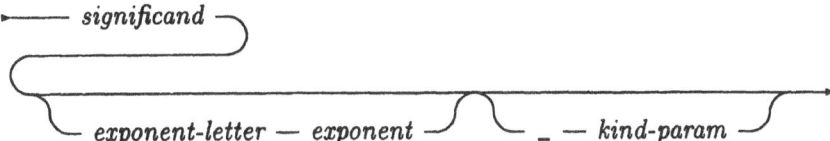

R414 *significand*

———— *digit-string* — . — *digit-string* ———►

R415 *exponent-letter*

———— E ———►

R416 *exponent*

———— *signed-digit-string* ———►

Constraint: The value of *kind-param* shall specify an approximation method that exists on the processor.

R417 *complex-literal-constant*

———— (— *real-part* — , — *imag-part* —) ———►

R418 *real-part*

———— *signed-real-literal-constant* ———►

R419 *imag-part*

———— *signed-real-literal-constant* ———►

Constraint: Both *real-part* and *imag-part* must either have no *kind-param* or have the same *kind-param*.

R420 *char-literal-constant*

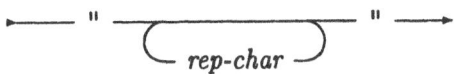

Note: Within a *char-literal-constant* a delimiter may be doubled to indicate a single instance of the delimiter.

R421 *logical-literal-constant*

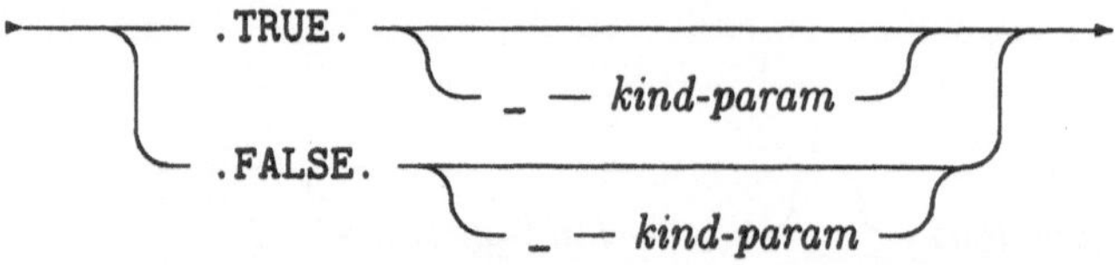

Constraint: The value of *kind-param* shall specify a representation method that exists on the processor.

Constraint: No integer, real, logical, or character literal constant, or *real-part* or *imag-part* shall be split onto more than one line via statement continuation.

R422 *derived-type-def*

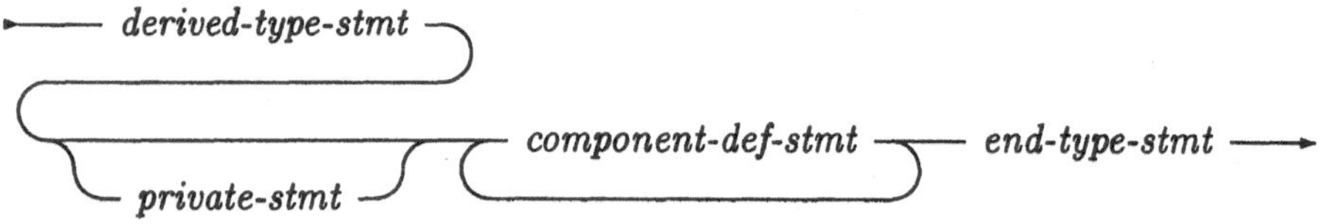

R423 *derived-type-stmt*

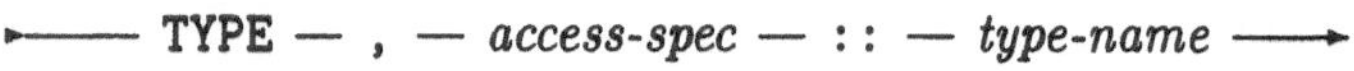

R424 *private-stmt*

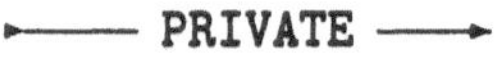

Constraint: A derived type *type-name* shall not be the same as the name of any intrinsic type nor the same as any other accessible derived type type-name.

R425 *component-def-stmt*

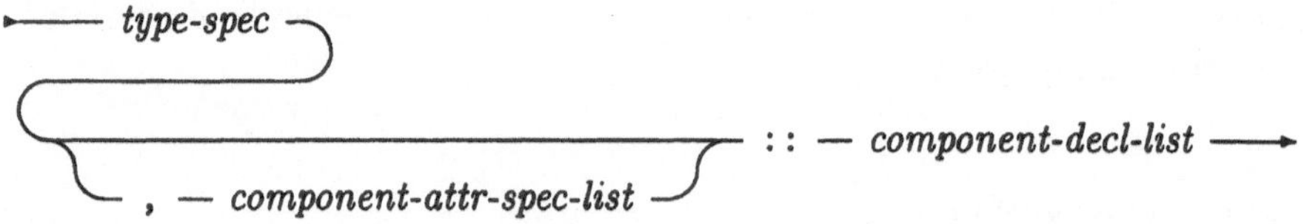

Constraint: The character length specified by the *char-len-param-value* in a *type-spec* shall be a constant specification expression.

R426 *component-attr-spec*

```
┌──────── POINTER ───────────────────────────────────────┐
●─┤                                                        ├──▶
  └── DIMENSION ── ( ── component-array-spec ── ) ──┘
```

R427 *component-array-spec*

```
┌──────── explicit-shape-spec-list ───────┐
●─┤                                         ├──▶
  └── deferred-shape-spec-list ──┘
```

Constraint: If a component of a derived type is of a type that is private, either the derived type definition shall contain the PRIVATE statement or the derived type shall be private.

Constraint: If a derived type is private it shall not contain a PRIVATE statement.

Constraint: No *component-attr-spec* shall appear more than once in a given *component-def-stmt*.

Constraint: If the POINTER attribute is not specified for a component, a *type-spec* in the *component-def-stmt* shall specify an intrinsic type or a previously defined derived type.

Constraint: If the POINTER attribute is specified for a component, a *type-spec* in the *component-def-stmt* shall specify an intrinsic type or any accessible derived type including the type being defined.

Constraint: If the POINTER attribute is specified, a *component-array-spec* shall be a *deferred-shape-spec-list*.

Constraint: If the POINTER attribute is not specified, each *component-array-spec* shall be an *explicit-shape-spec-list*.

Constraint: Each bound in the *explicit-shape-spec* shall be a constant specification expression.

R428 *component-decl*

```
●──── component-name ───▶
```

R430 *end-type-stmt*

```
●──── END TYPE ── type-name ───▶
```

Constraint: The *type-name* shall be the same as that in the corresponding *derived-type-stmt*.

R431 *structure-constructor*

►——— *type-name* — (— *expr-list* —) ——►

R432 *array-constructor*

►——— (/ — *ac-value-list* — /) ——►

R433 *ac-value*

►———⌐—— *expr* ———⌐——►
  └— *ac-implied-do* —┘

R434 *ac-implied-do*

►——— (— *ac-value-list* — , — *ac-implied-do-control* —) ——►

R435 *ac-implied-do-control*

►——— *ac-do-variable* — = —⌐
⌐————————————————————————┘
└— *scalar-int-expr* — , — *scalar-int-expr* ———————————————⌐——►
                            └— , — *scalar-int-expr* —┘

R436 *ac-do-variable*

►——— *scalar-int-variable* ——►

Constraint: An *ac-do-variable* shall be a named variable, shall not be a dummy argument, shall not have the POINTER attribute, shall not be initialized, shall not have the SAVE attribute and shall not be accessed by USE or host association, and shall be used in the scoping unit only as an *ac-do-variable*.

Constraint: Each *ac-value* expression in the *array-constructor* shall have the same type, kind type parameter, and character length.

E.5 Data Object Declarations and Specifications

R501 *type-declaration-stmt*

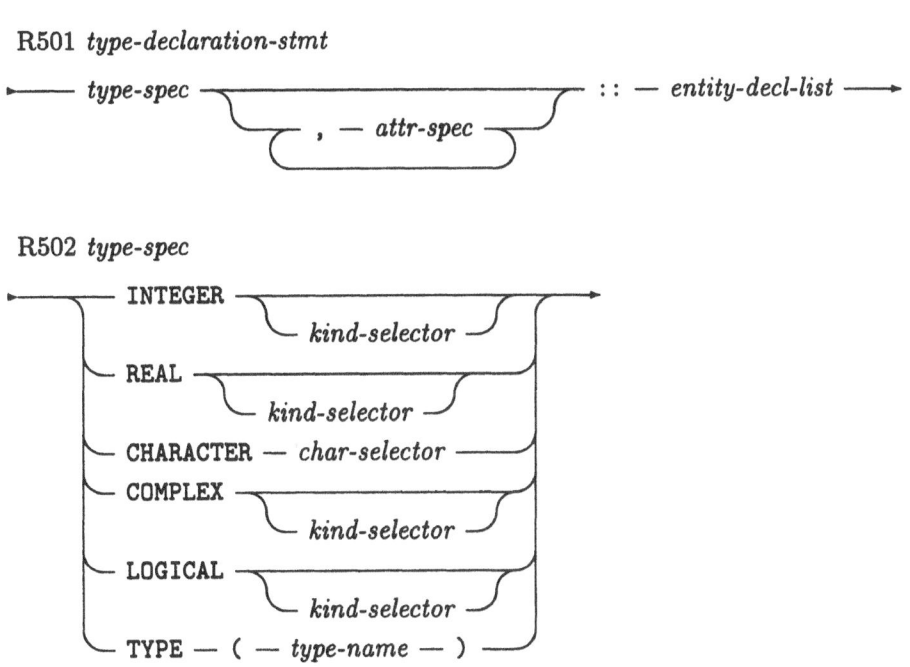

R502 *type-spec*

R503 *attr-spec*

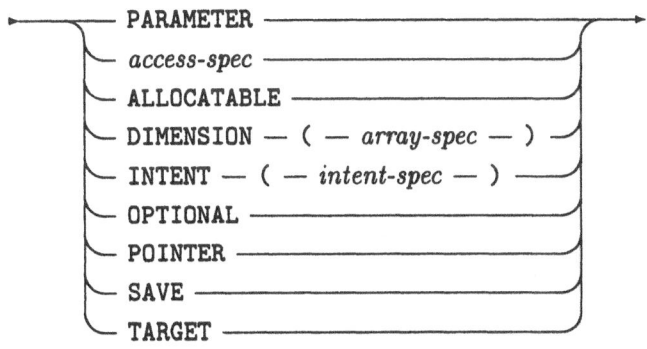

R504 *entity-decl*

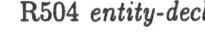

R505 *initialization*

►——— = — *initialization-expr* ——►

R506 *kind-selector*

►——— (— KIND — = — *scalar-int-constant-name* —) ——►

Constraint: The same *attr-spec* shall not appear more than once in a given *type-declaration-stmt*.

Constraint: The ALLOCATABLE attribute may be used only when declaring an array that is not a dummy argument or a function result.

Constraint: An array declared with a POINTER or an ALLOCATABLE attribute shall be specified with an *array-spec* that is a *deferred-shape-spec-list*.

Constraint: An *array-spec* for an *object-name* that is a function result that does not have the POINTER attribute shall be an *explicit-shape-spec-list*.

Constraint: If the POINTER attribute is specified, neither the TARGET nor INTENT attribute shall be specified.

Constraint: If the TARGET attribute is specified, neither the POINTER nor PARAMETER attribute shall be specified.

Constraint: The PARAMETER attribute shall not be specified for dummy arguments, pointers, allocatable arrays, or functions results.

Constraint: The INTENT and OPTIONAL attributes may be specified only for dummy arguments.

Constraint: An entity shall not have the PUBLIC attribute if its type has the PRIVATE attribute.

Constraint: The SAVE attribute shall not be specified for an object that is a dummy argument, a procedure, a function result, an automatic data object, or an object with the PARAMETER attribute.

Constraint: An array shall not have both the ALLOCATABLE attribute and the POINTER attribute.

Constraint: If the statement contains a PARAMETER attribute, *initialization* shall appear.

Constraint: If *object-name* is a dummy argument, a function result, an allocatable array, or an automatic object, *initialization* shall not appear.

Constraint: If *initialization* appears in a main program, the object shall have the PARAMETER attribute.

Constraint: If *initialization* appears, the object shall not have the POINTER attribute.

Constraint: The value of *scalar-int-constant-name* in *kind-selector* shall be nonnegative and shall specify a representation method that exists on the processor.

Constraint: If *initialization* appears, the statement shall contain either a PARAMETER attribute or a SAVE attribute.

R507 *char-selector*

────── (─ LEN ─ = ─ *char-len-param-value* ─) ──────→

R510 *char-len-param-value*

```
            ── specification-expr ──
─────────────<                      >────────→
            ──── * ──────────────────
```

Constraint: The *char-len-param-value* must be * for a dummy argument.

Constraint: The *char-len-param-value* may be * only for a dummy argument or a named constant.

R511 *access-spec*

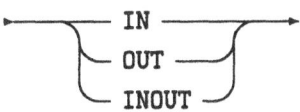

```
        ── PUBLIC ──
────────<            >────────→
        ── PRIVATE ──
```

Constraint: An *access-spec* shall appear only in the specification part of a module.

Constraint: An *access-spec* shall appear on every *type-declaration-statement* in the specification part of a module.

R512 *intent-spec*

```
        ── IN ────
────────<          >────────→
        ── OUT ───
        ── INOUT ─
```

Constraint: The INTENT attribute shall not be specified for a dummy argument that is a dummy procedure or a dummy pointer.

Constraint: A dummy argument with the INTENT(IN) attribute, or a subobject of such a dummy argument, shall not appear as

(1) The *variable* of an *assignment-stmt*,

(2) An *input-item* in a *read-stmt*,

(3) An *internal-file-unit* in a *write-stmt*,

(4) An IOSTAT= or SIZE= specifier in an input/output statement,

(5) A definable variable in an INQUIRE statement,

(6) A *stat-variable* or *allocate-object* in an *allocate-stmt* or a *deallocate-stmt*, or

(7) An actual argument in a reference to a procedure when the corresponding dummy argument has the INTENT(OUT) or INTENT(INOUT) attribute.

R513 *array-spec*

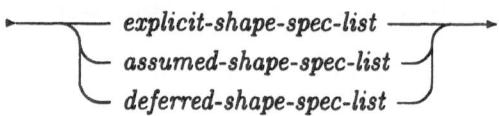

Constraint: The maximum rank is seven.

R514 *explicit-shape-spec*

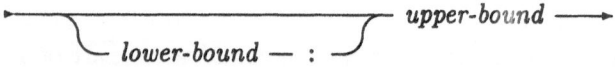

R515 *lower-bound*

——— specification-expr ———→

R516 *upper-bound*

——— specification-expr ———→

Constraint: An explicit-shape array whose bounds depend on the values of nonconstant expressions shall be a function result, or an automatic array of a procedure.

R517 *assumed-shape-spec*

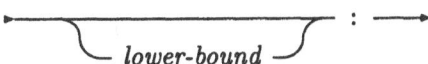

Constraint: All dummy argument arrays shall be assumed-shape arrays.

Constraint: Only dummy argument arrays shall be assumed-shape arrays.

R518 *deferred-shape-spec*

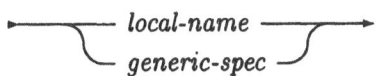

R522 *access-stmt*

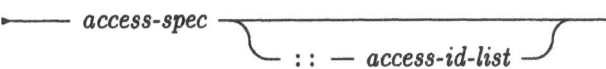

R523 *access-id*

Constraint: Each *local-name* shall be the name of a *generic-spec* of a *module-procedure-interface-block* or the name of a procedure that is not accessed by USE association. Each such *generic-spec* or procedure shall be named on an *access-stmt*.

Constraint: A module procedure that has a dummy argument or function result of a type that has PRIVATE accessibility shall have PRIVATE accessibility and shall not have a generic identifier that has PUBLIC accessibility.

R540 *implicit-stmt*

⊢――― IMPLICIT NONE ――→

Constraint: An *implicit-stmt* may appear only in the the specification part of a main program or private module.

E

E.6   Use of Data Objects

R601 *variable*

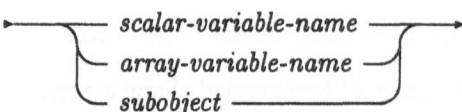

Constraint: *array-variable-name* shall be the name of a data object that is an array.

Constraint: *array-variable-name* shall not have the PARAMETER attribute.

Constraint: *scalar-variable-name* shall not have the PARAMETER attribute.

Constraint: *subobject* shall not be a subobject designator (for example, a substring) whose parent is a constant.

R602 *subobject*

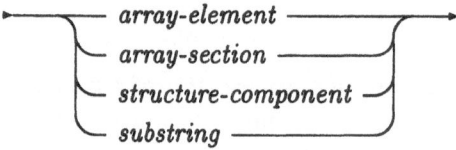

R603 *logical-variable*

►——— *variable* ——►

Constraint: *logical-variable* shall be of type logical.

R604 *default-logical-variable*

►——— *variable* ——►

Constraint: *default-logical-variable* shall be of type default logical.

R605 *char-variable*

►——— *variable* ——►

Constraint: *char-variable* shall be of type character.

R607 *int-variable*

►———— *variable* ————►

Constraint: *int-variable* shall be of type integer.

R608 *default-int-variable*

►———— *variable* ————►

Constraint: *default-int-variable* shall be of type default integer.

R609 *substring*

►——— *parent-string* — (— *substring-range* —) ——►

R610 *parent-string*

►┬——— *scalar-variable-name* ———┬——►
 ├——— *array-element* ———┤
 └——— *scalar-structure-component* —┘

R611 *substring-range*

►┬————————┬ : ┬————————┬——►
 └ *scalar-int-expr* ┘   └ *scalar-int-expr* ┘

Constraint: *parent-string* shall be of type character.

R612 *data-ref*

►——— *part-ref* ┬————————————┬——►
                └ % — *part-ref* ┘

R613 *part-ref*

►——— *part-name* ┬——————————————┬——►
                 └ (— *section-subscript-list* —) ┘

E

Constraint: In a *data-ref*, each *part-name* except the rightmost shall be of derived type.

Constraint: In a *data-ref*, each *part-name* except the leftmost shall be the name of a component of the derived type definition of the type of the preceding *part-name*.

Constraint: In a *part-ref* containing a *section-subscript-list*, the number of *section-subscripts* shall equal the rank of *part-name*.

Constraint: In a *data-ref*, there shall not be more than one *part-ref* with nonzero rank. A *part-name* to the right of a *part-ref* with nonzero rank shall not have the POINTER attribute.

R614 *structure-component*

→ *data-ref* →

Constraint: In a *structure-component*, there shall be more than one *part-ref* and the rightmost *part-ref* shall be of the form *part-name*.

R615 *array-element*

→ *data-ref* →

Constraint: In an *array-element*, every *part-ref* shall have rank zero and the last *part-ref* shall contain a *subscript-list*.

R616 *array-section*

→ *data-ref* ——————————→
　　　└ (— *substring-range* —) ┘

Constraint: In an *array-section*, exactly one *part-ref* shall have nonzero rank, and either the final *part-ref* shall have a *section-subscript-list* with nonzero rank or another *part-ref* shall have nonzero rank.

Constraint: In an *array-section* with a *substring-range*, the rightmost *part-name* shall be of type character.

R617 *subscript*

→ *scalar-int-expr* →

R618 *section-subscript*

```
┌─── subscript ────────┐
├─── subscript-triplet ─┤
└─── vector-subsript ──┘
```

R619 *subscript-triplet*

```
└─ subscript ─┘ : └─ subscript ─┘ └─ : ─ stride ─┘
```

R620 *stride*

────── *scalar-int-expr* ──────►

R621 *vector-subscript*

────── *int-expr* ──────►

Constraint: A *vector-subscript* shall be an integer array expression of rank one.

R622 *allocate-stmt*

```
───── ALLOCATE ──┐
                 │
└─ ( ─ allocation-list ──┬──────────────────────────┬── ) ──►
                         └─ , ─ STAT ─ = ─ stat-variable ─┘
```

R623 *stat-variable*

────── *scalar-int-variable* ──────►

R624 *allocation*

```
────── allocate-object ──┬──────────────────────────────────┬──►
                         └─ ( ─ allocate-shape-spec-list ─ ) ─┘
```

R625 *allocate-object*

```
┌─── variable-name ──────┐
└─── structure-component ─┘
```

R626 *allocate-shape-spec*

R627 *allocate-lower-bound*

▸─── *scalar-int-expr* ───▸

R628 *allocate-upper-bound*

▸─── *scalar-int-expr* ───▸

Constraint: Each *allocate-object* shall be a pointer or an allocatable array.

Constraint: The number of *allocate-shape-spec*s in an *allocate-shape-spec-list* shall be the same as the rank of the pointer or allocatable array.

R629 *nullify-stmt*

▸─── NULLIFY ─ (─ *pointer-object-list* ─) ───▸

R630 *pointer-object*

Constraint: Each *pointer-object* shall have the POINTER attribute.

R631 *deallocate-stmt*

Constraint: Each *allocate-object* shall be a pointer or allocatable array.

E.7 Expressions and Assignment

R701 *primary*

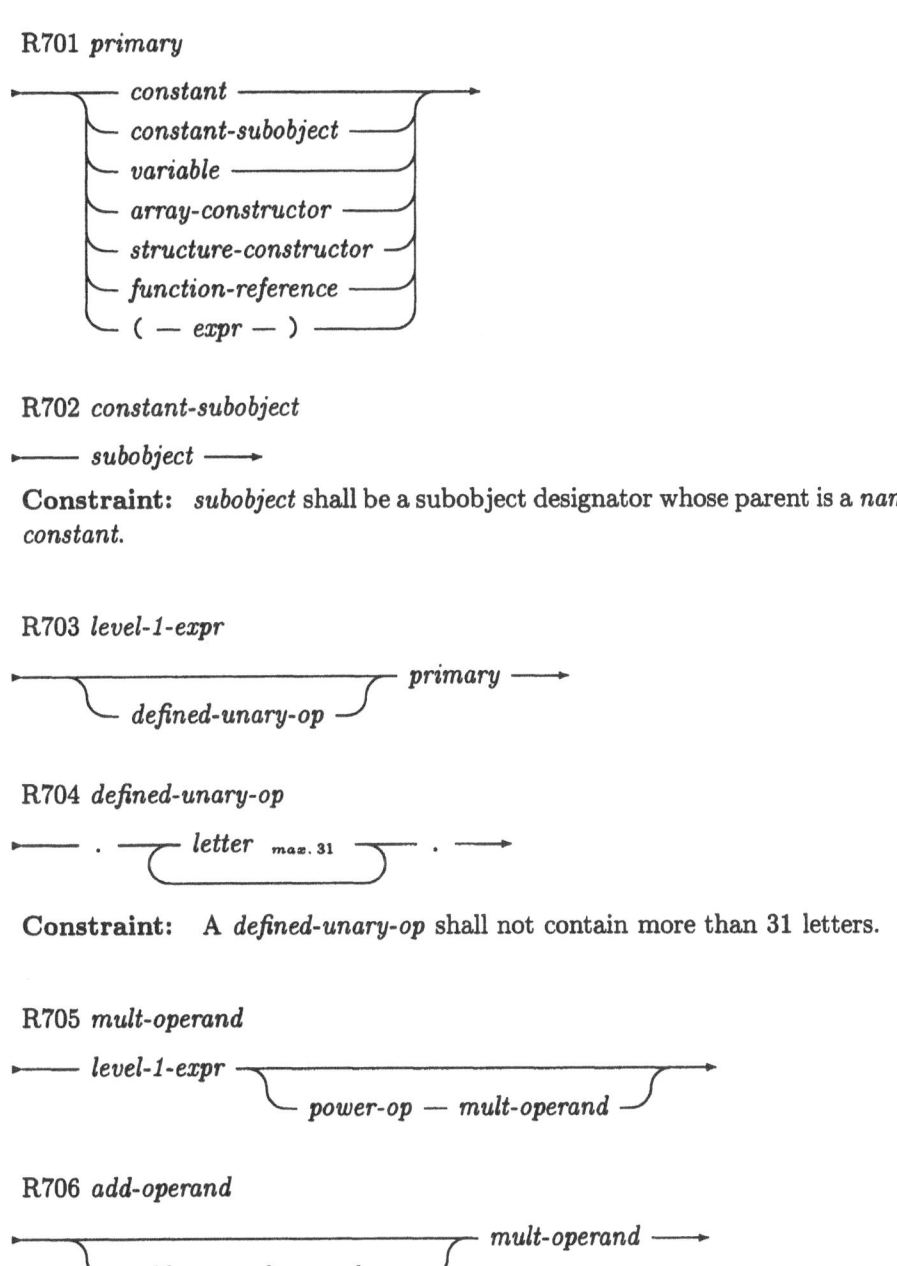

R702 *constant-subobject*

►———— *subobject* ————►

Constraint: *subobject* shall be a subobject designator whose parent is a *named-constant*.

R703 *level-1-expr*

R704 *defined-unary-op*

Constraint: A *defined-unary-op* shall not contain more than 31 letters.

R705 *mult-operand*

R706 *add-operand*

R707 *level-2-expr*

R708 *power-op*

$$\longmapsto \ ** \longrightarrow$$

R709 *mult-op*

R710 *add-op*

R711 *level-3-expr*

R712 *concat-op*

$$\longmapsto \ // \longrightarrow$$

R713 *level-4-expr*

R714 *rel-op*

R715 *and-operand*

```
┌─────────────────────┐ level-4-expr ───►
└─ not-op ─┘
```

R716 *or-operand*

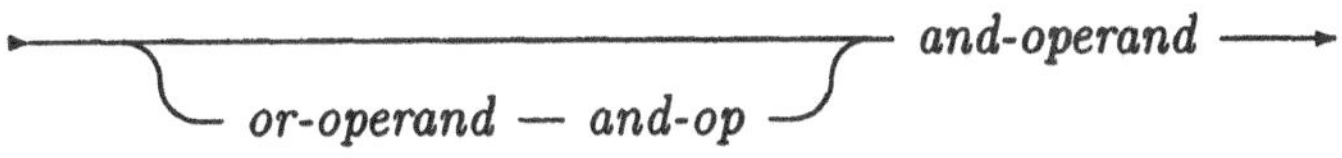

```
┌──────────────────────────────┐ and-operand ───►
└─ or-operand ─ and-op ─┘
```

R717 *equiv-operand*

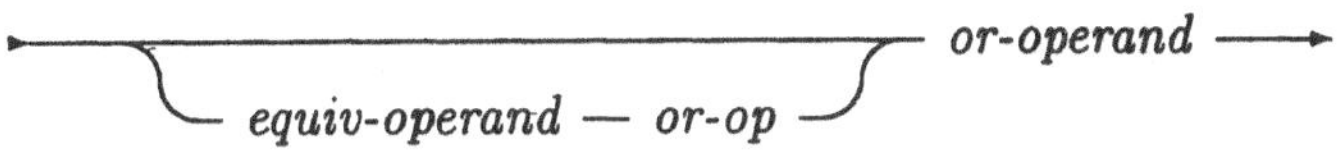

```
┌──────────────────────────────┐ or-operand ───►
└─ equiv-operand ─ or-op ─┘
```

R718 *level-5-expr*

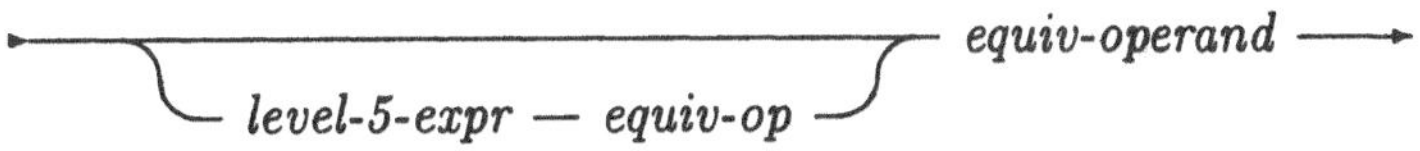

```
┌──────────────────────────────────┐ equiv-operand ───►
└─ level-5-expr ─ equiv-op ─┘
```

R719 *not-op*

```
►─── .NOT. ───►
```

R720 *and-op*

```
►─── .AND. ───►
```

R721 *or-op*

```
►─── .OR. ───►
```

R722 *equiv-op*

```
►─┬─ .EQV. ─┬─►
  └─ .NEQV. ─┘
```

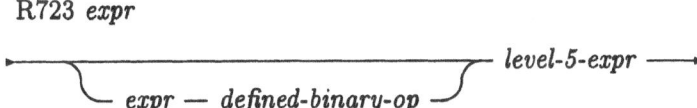

R723 *expr*

```
┌──────────────────────────────┐ level-5-expr ───►
└─ expr ─ defined-binary-op ─┘
```

R724 *defined-binary-op*

──── . ──⌐─ *letter* ₘₐₓ. 31 ─⌐── . ──►

Constraint: A *defined-binary-op* shall not contain more than 31 letters.

R725 *logical-expr*

──── *expr* ──►

Constraint: *logical-expr* shall be of type logical.

R726 *char-expr*

──── *expr* ──►

Constraint: *char-expr* shall of be type character.

R728 *int-expr*

──── *expr* ──►

Constraint: *int-expr* shall be of type integer.

R729 *numeric-expr*

──── *expr* ──►

Constraint: *numeric-expr* shall be of type integer, real or complex.

R730 *initialization-expr*

──── *expr* ──►

Constraint: *initialization-expr* shall be an initialization expression.

R731 *char-initialization-expr*

──── *char-expr* ──►

Constraint: *char-initialization-expr* shall be an initialization expression.

R732 *int-initialization-expr*

──── *int-expr* ──►

Constraint: *int-initialization-expr* shall be an initialization expression.

R733 *logical-initialization-expr*

▸───── *logical-expr* ────▸

Constraint: *logical-initialization-expr* shall be an initialization expression.

R734 *specification-expr*

▸───── *scalar-int-expr* ────▸

Constraint: The *scalar-int-expr* shall be a restricted expression.

R735 *assignment-stmt*

▸───── *variable* ─ **=** ─ *expr* ────▸

R736 *pointer-assignment-stmt*

▸───── *pointer-object* ─ **=>** ─ *target* ────▸

R737 *target*

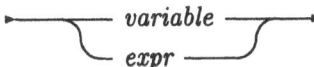

Constraint: The *pointer-object* shall have the POINTER attribute.

Constraint: The *variable* shall have the TARGET attribute or be a subobject of an object with the TARGET attribute, or it shall have the POINTER attribute.

Constraint: The *target* shall be of the same type, type parameters, and rank as the pointer.

Constraint: The *target* shall not be an array with a vector subscript.

Constraint: The *expr* shall deliver a pointer result.

E

R739 *where-construct*

```
┌──── where-construct-stmt ──────┐
│                    └─ assignment-stmt ─┘ │
├─────────────────────────────────────┬─ end-where-stmt ──→
  └─ elsewhere-stmt ──┬──────────────┘
                      └─ assignment-stmt ─┘
```

R740 *where-construct-stmt*

──── WHERE ─ (─ *mask-expr* ─) ──→

R743 *mask-expr*

──── *logical-expr* ──→

R745 *elsewhere-stmt*

──── ELSEWHERE ──→

R746 *end-where-stmt*

──── END WHERE ──→

Constraint: In each *assignment-stmt*, the *mask-expr* and the *variable* being defined must be arrays of the same shape.

Constraint: The *assignment-stmt* must not be a defined assignment.

E.8 Execution Control

R801 *block*

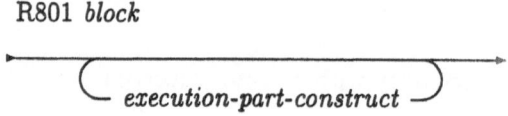

R802 *if-construct*

```
▸──── if-then-stmt ── block ──┬─────────────────────────────┐
                              └── else-if-stmt ── block ──┘
```

```
┌──────────────────────── end-if-stmt ───▸
└── else-stmt ── block ──┘
```

R803 *if-then-stmt*

▸──── IF ── (── *scalar-logical-expr* ──) ── THEN ───▸

R804 *else-if-stmt*

▸──── ELSE IF ── (── *scalar-logical-expr* ──) ── THEN ───▸

R805 *else-stmt*

▸──── ELSE ───▸

R806 *end-if-stmt*

▸──── END IF ───▸

R808 *case-construct*

```
▸──── select-case-stmt ──┬─────────────────────────┐
                         └── case-stmt ── block ──┘
```

```
┌──────────────────────── end-select-stmt ───▸
└── CASE DEFAULT ── block ──┘
```

R809 *select-case-stmt*

▸──── SELECT CASE ── (── *case-expr* ──) ───▸

R810 *case-stmt*

▸──── CASE ── *case-selector* ───▸

E

R811 *end-select-stmt*

►——— END SELECT ———►

R812 *case-expr*

►——┬——— scalar-int-expr ———┬——►
    └——— scalar-char-expr ———┘

R813 *case-selector*

►——— (— case-value-range-list —) ———►

R814 *case-value-range*

►——┬——— case-value ————————————┬——►
    ├——— case-value — : ————————┤
    ├——— : — case-value ————————┤
    └——— case-value — : — case-value ——┘

R815 *case-value*

►——┬——— scalar-int-initialization-expr ———┬——►
    └——— scalar-char-initialization-expr ———┘

Constraint: For a given *case-construct*, each *case-value* shall be of the same type as *case-expr*. For character type, length differences are allowed.

Constraint: For a given *case-construct*, the *case-value-ranges* shall not overlap; that is, there shall be no possible value of the *case-expr* that matches more than one *case-value-range*.

R816 *do-construct*

►——— block-do-construct ———►

R817 *block-do-construct*

►——— do-stmt — do-block — end-do ———►

R818 *do-stmt*

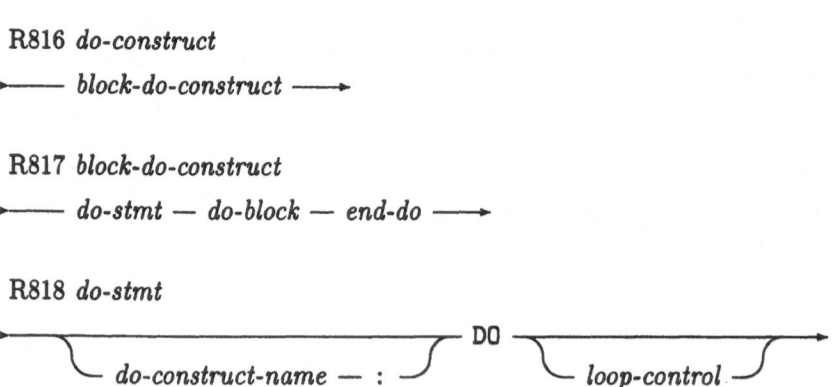

Constraint: The *do-construct-name* shall not be the same as the name of any accessible entity.

Constraint: The same *do-construct-name* shall not be used on more than one *do-stmt* in a scoping unit.

R821 *loop-control*

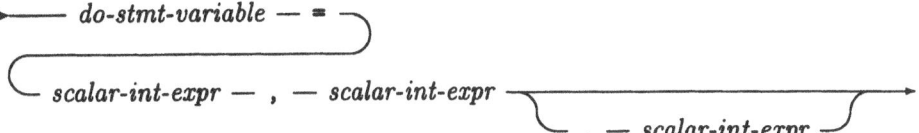

R822 *do-variable*

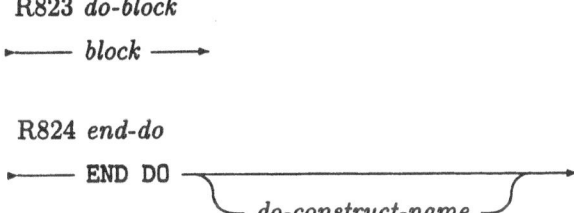

Constraint: A *do-stmt-variable* shall be a named variable, shall not be a dummy argument, shall not have the POINTER attribute, and shall not be accessed by USE or host association.

R823 *do-block*

►──── *block* ──→

R824 *end-do*

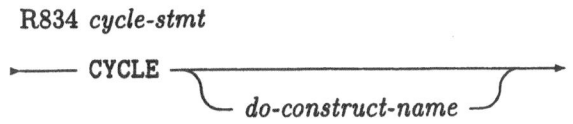

Constraint: If the *do-stmt* is identified by a *do-construct-name*, the corresponding *end-do* shall specify the same *do-construct-name*. If the *do-stmt* is not identified by a *do-construct-name*, the corresponding *end-do* shall not specify a *do-construct-name*.

R834 *cycle-stmt*

►──── CYCLE ────────────────────→
           └── *do-construct-name* ──┘

Constraint: If a *cycle-stmt* refers to a *do-construct-name*, it shall be within the range of that *do-construct*; otherwise, it shall be within the range of at least one *do-construct*.

R835 *exit-stmt*

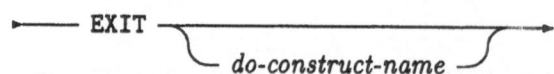

Constraint: If an *exit-stmt* refers to a *do-construct-name*, it shall be within the range of that *do-construct*; otherwise, it shall be within the range of at least one *do-construct*.

R840 *stop-stmt*

►——— STOP ——►

E.9   Input/Output Statements

R901 *io-unit*

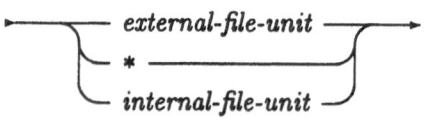

R902 *external-file-unit*

►——— *scalar-int-expr* ——►

R903 *internal-file-unit*

►——— *char-variable* ——►

Constraint: The *char-variable* shall not be an array section with a vector subscript.

R904 *open-stmt*

►——— OPEN — (— *connect-spec-list* —) ——►

R905 *connect-spec*

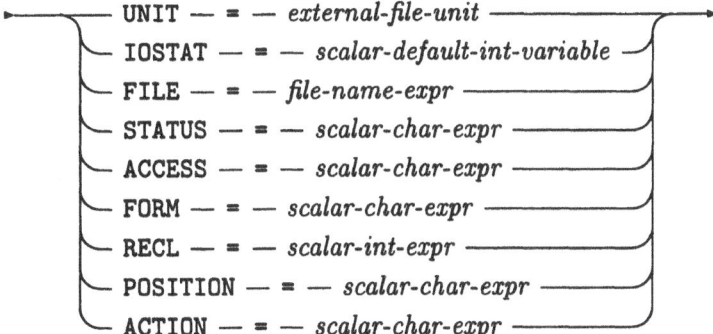

R906 *file-name-expr*

▸─────── *scalar-char-expr* ───▸

Constraint: A *connect-spec-list* shall contain exactly one UNIT = *io-unit*, exactly one STATUS = *scalar-char-expr*, and exactly one ACTION = *scalar-char-expr* and may contain at most one of each of the other specifiers.

R907 *close-stmt*

▸──── CLOSE — (— *close-spec-list* —) ───▸

R908 *close-spec*

▸──────── UNIT — = — *external-file-unit* ───────▸
    └─ IOSTAT — = — *scalar-default-int-variable* ─┘
    └─ STATUS — = — *scalar-char-expr* ─────────┘

Constraint: A *close-spec-list* shall contain exactly one UNIT = *io-unit* and may contain at most one of each of the other specifiers.

R909 *read-stmt*

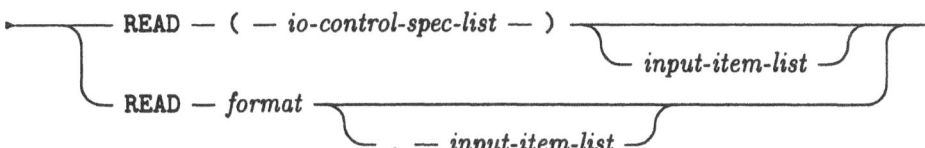

R910 *write-stmt*

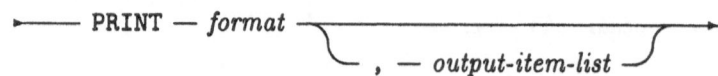

R911 *print-stmt*

PRINT — *format*
, — *output-item-list*

R912 *io-control-spec*

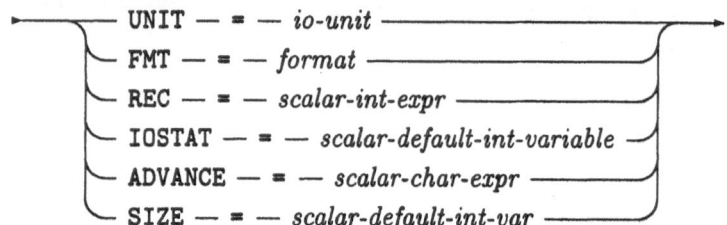

Constraint: An *io-control-spec-list* shall contain exactly one UNIT = *io-unit* and may contain at most one of each of the other specifiers.

Constraint: A SIZE= specifier shall not appear in a *write-stmt*.

Constraint: If the unit specifier specifies an internal file, the *io-control-spec-list* shall not contain a REC= specifier.

Constraint: If the REC= specifier is present, the *format*, if any, shall not be an asterisk specifying list-directed input/output.

Constraint: An ADVANCE= specifier may be present only in a formatted sequential input/output statement with explicit format specification whose control information list does not contain an internal file unit specifier.

Constraint: If a SIZE= specifier is present, an ADVANCE= specifier also shall appear.

R913 *format*

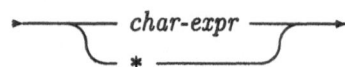

R914 *input-item*

▸—— *variable* ——▸

R915 *output-item*

▸—— *expr* ——▸

R919 *backspace-stmt*

▸—— BACKSPACE — (— *position-spec-list* —) ——▸

R920 *endfile-stmt*

▸—— ENDFILE — (— *position-spec-list* —) ——▸

R921 *rewind-stmt*

▸—— REWIND — (— *position-spec-list* —) ——▸

R922 *position-spec*

▸————— UNIT — = — *external-file-unit* ——————————▸
    └— IOSTAT — = — *scalar-default-int-variable* —┘

Constraint: A *position-spec-list* shall contain exactly one UNIT = *external-file-unit*, and may contain at most one IOSTAT specifier.

R923 *inquire-stmt*

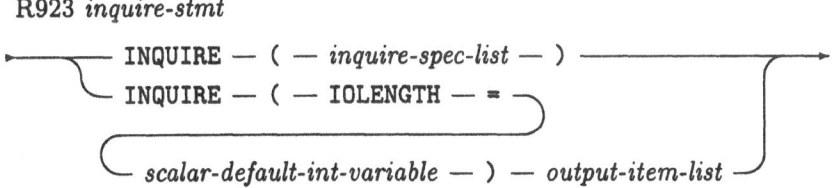

R924 *inquire-spec*

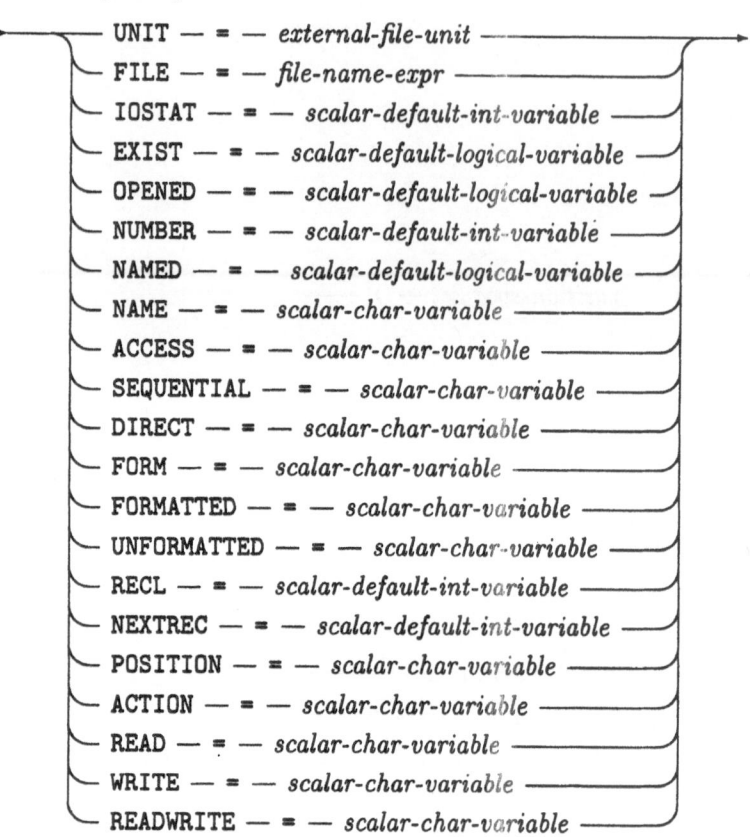

Constraint: An *inquire-spec-list* shall contain one FILE= specifier or one UNIT= specifier, but not both, and at most one of each of the other specifiers.

E.10   Input/Output Editing

R1002 *format-specification*

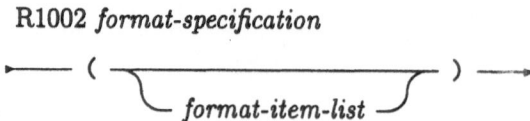

R1003 *format-item*

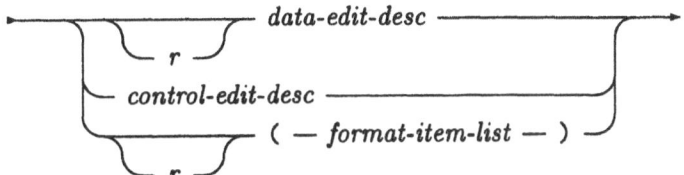

R1004 *r*

►——— *int-literal-constant* ——►

Constraint: *r* shall be positive.

Constraint: *r* shall not have a kind parameter specified for it.

R1005 *data-edit-desc*

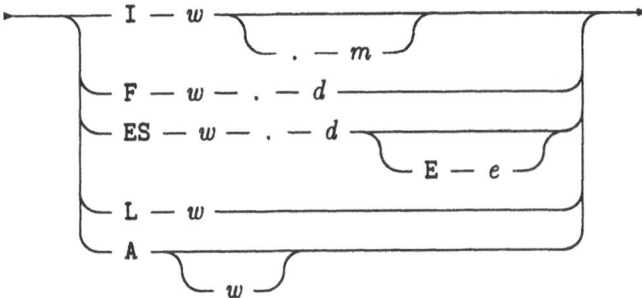

R1006 *w*

►——— *int-literal-constant* ——►

R1007 *m*

►——— *int-literal-constant* ——►

R1008 *d*

►——— *int-literal-constant* ——►

R1009 *e*

►——— *int-literal-constant* ——►

Constraint: *w* and *e* shall be positive.

Constraint: *w*, *m*, *d*, and *e* shall not have kind parameters specified for them.

R1010 *control-edit-desc*

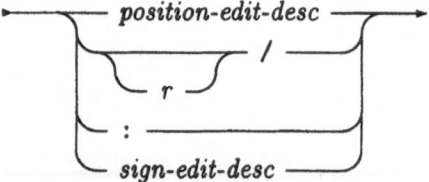

R1012 *position-edit-desc*

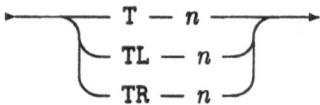

R1013 *n*

▶──── *int-literal-constant* ────▶

Constraint: *n* shall be positive.

Constraint: *n* shall not have a kind parameter specified for it.

R1014 *sign-edit-desc*

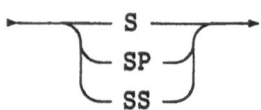

E.11 Program Units

R1101 *main-program*

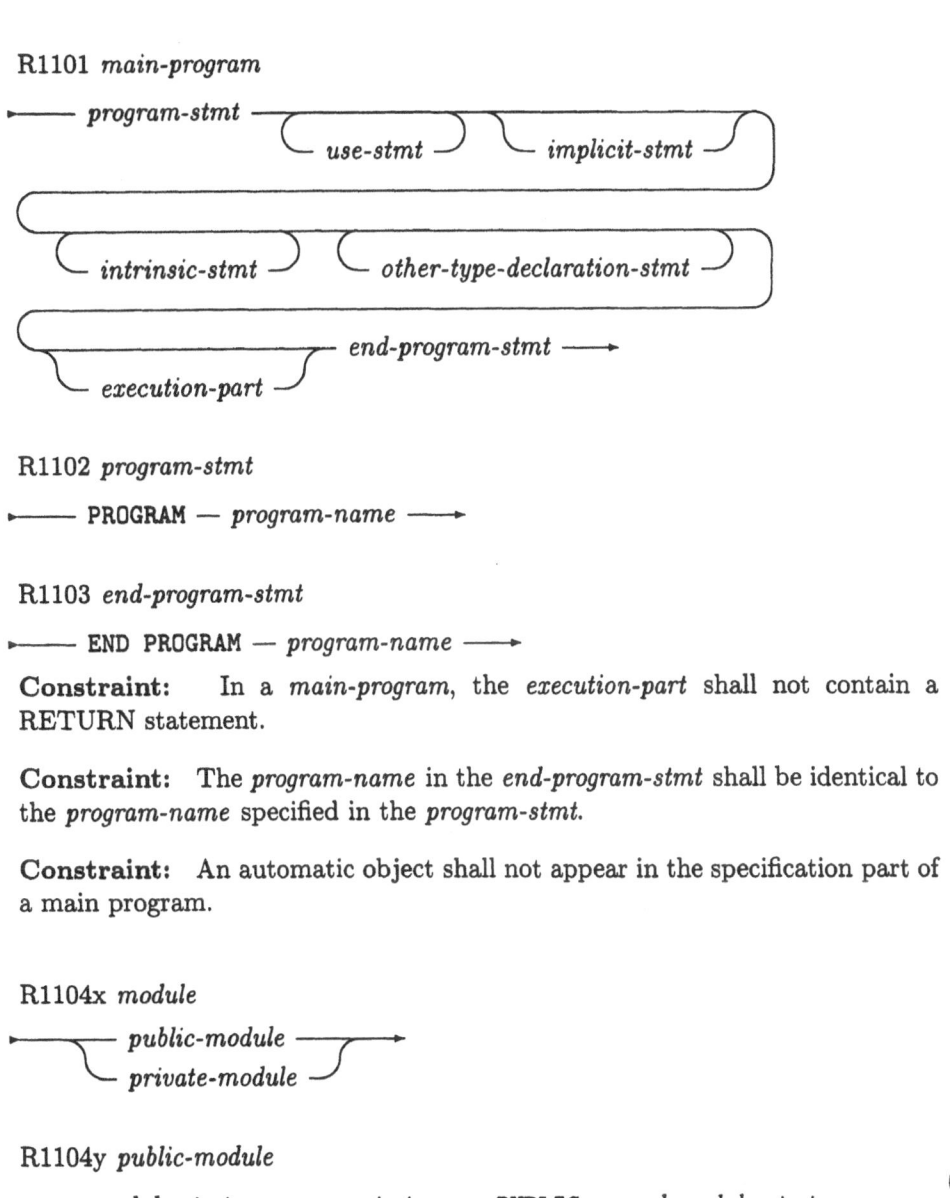

R1102 *program-stmt*

———— PROGRAM — *program-name* ——►

R1103 *end-program-stmt*

———— END PROGRAM — *program-name* ——►

Constraint: In a *main-program*, the *execution-part* shall not contain a RETURN statement.

Constraint: The *program-name* in the *end-program-stmt* shall be identical to the *program-name* specified in the *program-stmt*.

Constraint: An automatic object shall not appear in the specification part of a main program.

R1104x *module*

———— *public-module* ————►
—— *private-module* ——

R1104y *public-module*

———— *module-stmt* ——— *use-stmt* ——— PUBLIC — *end-module-stmt* ——►

E

R1104 *private-module*

Constraint: A PRIVATE statement shall appear if any *use-stmt*s appear. A PRIVATE statement shall not appear if no *use-stmt*s are present.

R1104z *module-entity-def*

- *derived-type-def*
- *other-type-declaration-stmt*
- *module-procedure-interface-block*
- *external-procedure-interface-block*

R212 *module-subprogram-part*

— *contains-stmt* — *module-subprogram* —

R213 *module-subprogram*

- *function-subprogram*
- *subroutine-subprogram*

Constraint: Every *function-subprogram* or *subroutine-subprogram* in a *private-module* shall be listed in an *access-stmt*.

R1105 *module-stmt*

— MODULE — *module-name* —→

R1106 *end-module-stmt*

— END MODULE — *module-name* —→

Constraint: The *module-name* is specified in the *end-module-stmt* shall be identical to the *module-name* specified in the *module-stmt*.

Constraint: An automatic object shall not appear in the specification part of a module.

R1107 *use-stmt*

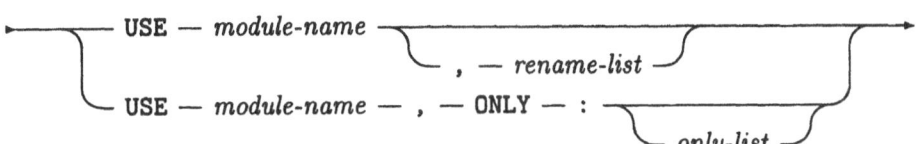

Constraint: A *module-name* shall appear in at most one USE statement in a scoping unit.

Constraint: A module shall not be made accessible by more than one use statement in a scoping unit.

Constraint: The module shall appear in a previously processed program unit.

Constraint: There shall be at least one *only* in the *only-list*.

R1108 *rename*

►——— *local-name* — => — *use-name* ———►

R1109 *only*

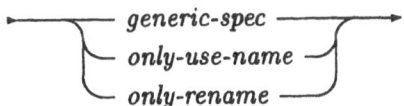

R1110 *only-use-name*

►——— *use-name* ———►

R1111 *only-rename*

►——— *local-name* — => — *use-name* ———►

Constraint: Each *generic-spec* shall be a public entity in the module.

Constraint: Each *use-name* shall be the name of a public entity in the module.

Constraint: *use-name* shall not be the name of an intrinsic procedure.

Constraint: In a *use-stmt* a *use-name* shall appear only once.

Constraint: No two accessible entities may have the same local name.

E.12   Procedures

R1201 *module-procedure-interface-block*

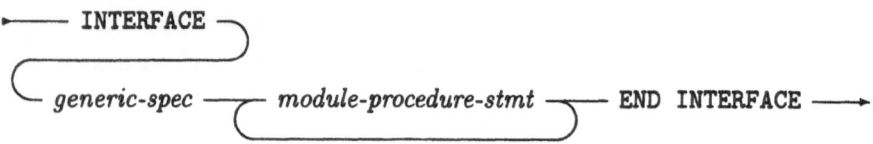

Constraint: Every *generic-spec* in a *private-module* shall be listed in an *access-stmt*.

Constraint: If *generic-spec* is also the name of an intrinsic procedure the generic name shall appear in a previous intrinsic statement in the module.

Constraint: If *generic-spec* is the same as the name of an intrinsic procedure the arguments to *generic-spec* must differ from the arguments to the intrinsic procedure in a way that allows unambiguous reference.

Constraint: *generic-spec* shall not be the same name as that of any accessible procedure or variable.

R1206 *module-procedure-stmt*

▶────── MODULE PROCEDURE — *procedure-name-list* ───▶

Constraint: A *procedure-name* in a *module-procedure-stmt* shall not be one which previously had been specified in any *module-procedure-stmt* with the same generic identifier in the same specification part.

Constraint: Each *procedure-name* must be accessible as a module procedure.

R1207 *generic-spec*

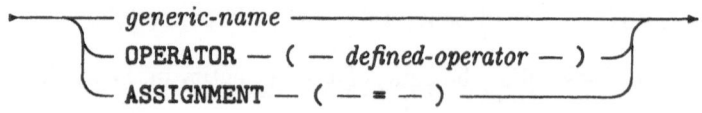

Constraint: *generic-name* shall not be the same as any module procedure name.

R1202 *dummy-procedure-interface-block*

▶──── INTERFACE ──┬── *interface-body* ──┬── END INTERFACE ────▶
                  └──────────────────┘

Constraint: Each procedure dummy argument shall appear in exactly one interface body.

Constraint: Each procedure specified shall be a dummy argument.

R1202x *external-procedure-interface-block*

▶──── INTERFACE ──┬── *interface-body* ──┬── END INTERFACE ────▶
                  └──────────────────┘

Constraint: The name of an external procedure shall not be the name of an accessible module procedure.

Constraint: An external procedure shall not be used as an actual argument.

Constraint: The *interface-body* of a dummy or external procedure shall specify the intents of all dummy arguments except pointer and procedure arguments.

R1205 *interface-body*

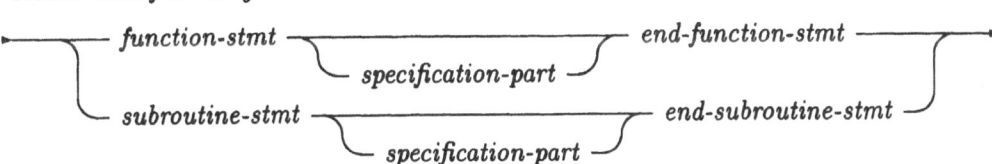

R1209 *intrinsic-stmt*

▶──── INTRINSIC ── *intrinsic-procedure-name-list* ────▶

Constraint: Each *intrinsic-procedure-name* shall be the name of an intrinsic procedure.

Ⓔ

Constraint: The specification part of a *dummy-procedure-interface-block* or *external-procedure-interface-block* shall not contain a *type-declaration-stmt* or interface block for a variable or procedure which is not a dummy argument or function result.

R1298 *intrinsic-procedure-name*

Syntactic class *intrinsic-procedure-name* represents one of the following terminal symbols:

abs	acos	adjustl	adjustr
aimag	aint	all	allocated
anint	any	asin	associated
atan	atan2	bit_size	btest
ceiling	char	cmplx	conjg
cos	cosh	count	cshift
date_and_time	digits	dot_product	eoshift
epsilon	exp	exponent	floor
fraction	huge	iand	ibclr
ibits	ibset	ichar	ieor
index	int	ior	ishft
ishftc	kind	lbound	len
len_trim	log	log10	logical
matmul	max	maxexponent	maxloc
maxval	merge	min	minexponent
minloc	minval	modulo	mvbits
nearest	nint	not	pack
precision	present	product	radix
random_number	random_seed	range	real
repeat	reshape	rrspacing	scale
scan	selected_int_kind	selected_real_kind	set_exponent
shape	sign	sin	sinh
size	spacing	spread	sqrt
sum	system_clock	tan	tanh
tiny	transpose	trim	ubound
unpack	verify		

R1299 *unsupported-intrinsic-name*

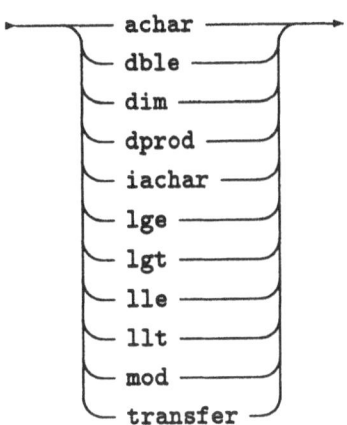

Constraint: In a reference to any intrinsic function that has a kind argument the associated actual argument must be a named constant.

R1210 *function-reference*

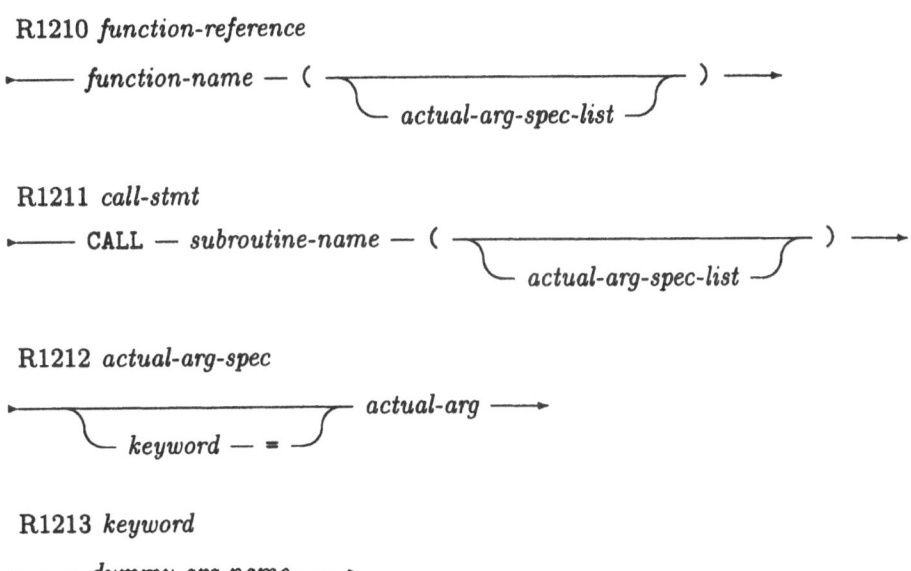

R1211 *call-stmt*

R1212 *actual-arg-spec*

R1213 *keyword*

▶──── *dummy-arg-name* ───▶

R1214 *actual-arg*

Constraint: The *keyword* = may be omitted from an *actual-arg-spec* only if the *keyword* = has been omitted from each preceding actual-arg-spec in the argument list.

Constraint: Each *keyword* shall be the name of a dummy argument of the procedure.

Constraint: In a reference to a function, a *procedure-name actual-arg* shall be the name of a function.

Constraint: A *procedure-name actual-arg* shall not be the name of an intrinsic procedure nor a *generic-name*.

R1216 *function-subprogram*

R1221 *subroutine-subprogram*

R1221v *specification-part*

R1221w *dummy-declaration-stmt*

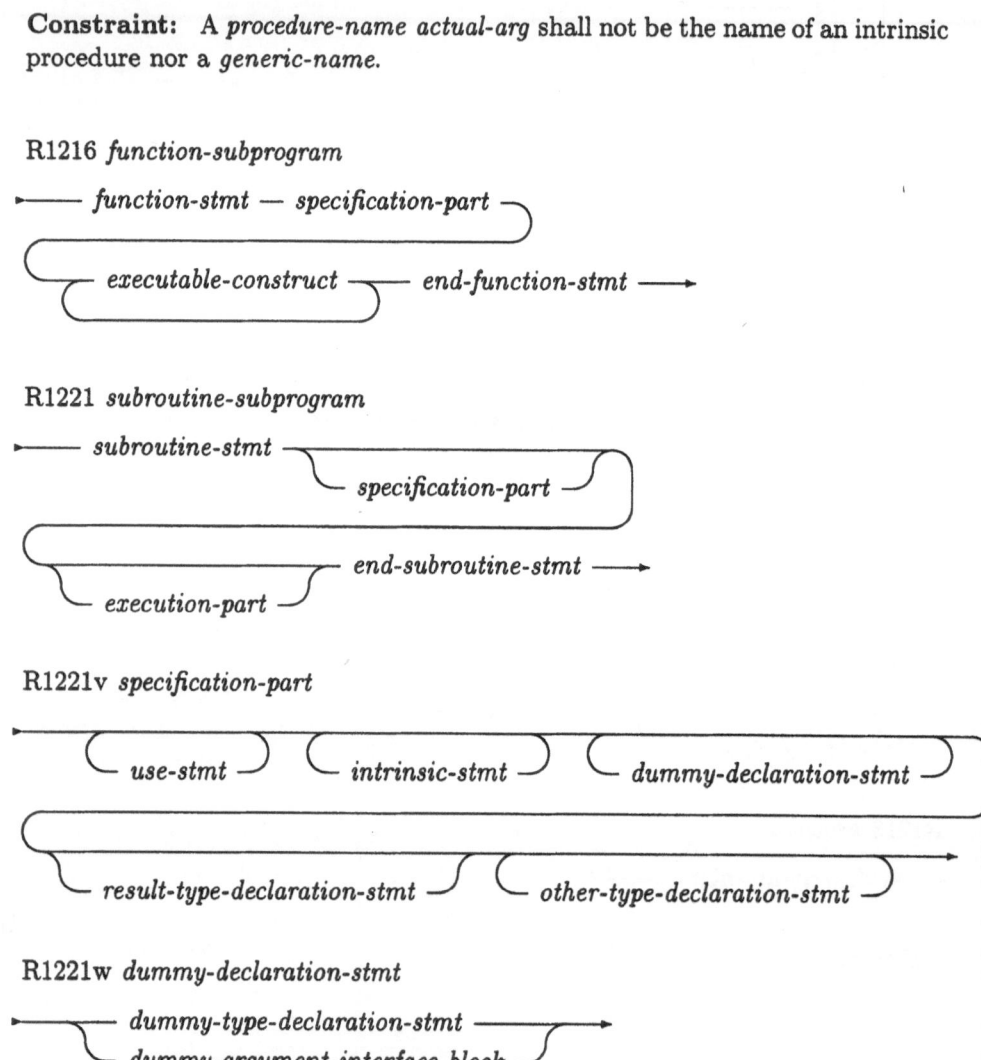

Constraint: A *result-type-declaration-stmt* shall appear in the *specification-part* of a function.

Constraint: A *result-type-declaration-stmt* shall not appear in the *specification-part* of a subroutine.

R1221x *dummy-type-declaration-stmt*

▶──── *type-declaration-stmt* ────▶

Constraint: Each entity in the *entity-decl-list* of a *dummy-type-declaration-stmt* shall be a dummy argument.

R1221y *result-type-declaration-stmt*

▶──── *type-declaration-stmt* ────▶

Constraint: There shall be exactly one entity in the *entity-decl-list* of a *result-type-declaration-stmt* and it shall be the function result.

R1221z *other-type-declaration-stmt*

▶──── *type-declaration-stmt* ────▶

Constraint: No entity in the *entity-decl-list* of an *other-type-declaration-stmt* shall be a dummy argument or a function result.

R1217 *function-stmt*

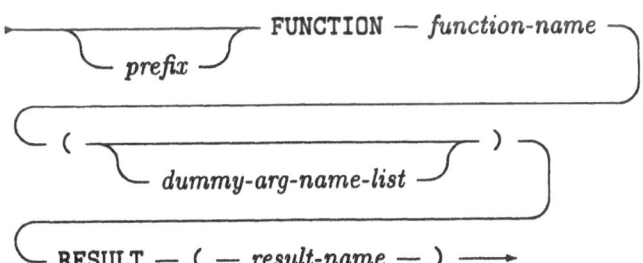

Constraint: The *function-name* shall not appear in any specification statement in the scoping unit of the function subprogram.

R1218 *prefix*

▶──── RECURSIVE ────▶

R1220 *end-function-stmt*

●────── END FUNCTION — *function-name* ────➤

Constraint: *result-name* shall not be the same as *function-name*.

Constraint: The *function-name* in the *end-function-stmt* shall be identical to the *function-name* specified in the *function-stmt*.

R1222 *subroutine-stmt*

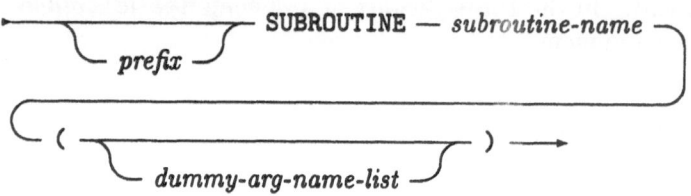

R1223 *dummy-arg*

●────── *dummy-arg-name* ────➤

Constraint: If a *dummy-arg* has the OPTIONAL attribute all subsequent *dummy-args* in *dummy-arg-name-list* shall also have the OPTIONAL attribute.

R1224 *end-subroutine-stmt*

●────── END SUBROUTINE — *subroutine-name* ────➤

Constraint: The *subroutine-name* in the *end-subroutine-stmt* shall be identical to the *subroutine-name* specified in the *subroutine-stmt*.

R1226 *return-stmt*

●────── RETURN ────➤

Constraint: The *return-stmt* shall be in the scoping unit of a function or subroutine subprogram.

R1227 *contains-stmt*

●────── CONTAINS ────➤

Constraint: A local variable declared in the specification part of a function shall not have the SAVE attribute (hence also cannot be initialized).

Constraint: The *specification-part* of a function subprogram shall specify that all dummy arguments have INTENT(IN) except procedure arguments and arguments with the POINTER attribute.

Constraint: The *specification-part* of a subroutine shall specify the intents of all dummy arguments except procedure arguments and arguments with the POINTER attribute.

Constraint: In a function any variable which is accessed by host or USE association, or is a dummy argument to a function shall not be used in the following contexts:

(1) As the *variable* of an *assignment-stmt*;

(2) As an *input-item* in a *read-stmt*;

(3) As an *internal-file-unit* in a *write-stmt*;

(4) As an IOSTAT= specifier in an input or output statement;

(5) As the *pointer-object* of a *pointer-assignment-stmt*;

(6) As the *target* of a *pointer-assignment-stmt*;

(7) As the *expr* of an *assignment-stmt* in which the *variable* is of a derived type if the derived type has a pointer component at any level of component selection;

(8) As an *allocate-object* or *stat-variable* in an *allocate-stmt* or *deallocate-stmt*, or as a *pointer-object* in a *nullify-stmt*; or

(9) As an actual argument associated with a dummy argument with the POINTER attribute.

Constraint: Any subprogram referenced in a function including one referenced via a defined operation, shall be a function or shall be referenced by defined assignment.

Constraint: Any subroutine referenced by defined assignment from a function, and any subprogram invoked during such reference, shall obey all of the constraints above relating to variables in a function except that the first argument to the subroutine may have INTENT(OUT) or INTENT(INOUT).

Constraint: A function shall not contain an *open-stmt*, *close-stmt*, *backspace-stmt*, *endfile-stmt*, *rewind-stmt*, or *inquire-stmt*.

Constraint: A function shall not contain a *read-stmt* or *write-stmt* unless it refers to an internal file, or have UNIT = * and have a FMT= specifier, or is a *read-stmt* with no *io-control-spec-list*.

INDEX

A edit descriptor 12-6
ABS function 14-2, 14-8
abstract data type 13-7
access method 11-3
ACCESS= specifier 11-31, 11-33, 11-35
accessible
    → host association 4-12
    → USE association 13-3
ACOS function 14-2, 14-8
ACTION= specifier 11-31, 11-35
actual argument list 13-28
ADJUSTL function 14-2, 14-8
ADJUSTR function 14-2, 14-8
ADVANCE= specifier 11-9
advancing input/output
AIMAG function 14-2, 14-8
AINT function 14-2, 14-9
ALL function 14-2, 14-9
allocatable array 4-6, 6-13
ALLOCATABLE attribute 9-3
ALLOCATE statement 5-1, 5-2, 6-14
ALLOCATED function 14-2, 14-9
allocation status 6-14
ANINT function 14-2, 14-9
ANY function 14-2, 14-9
argument association 4-12, 13-27, 13-29
argument keyword 13-29, 14-6
argument list 13-27
arithmetic → numeric
array 4-1, 4-5
array assignment statement 6-21, 8-1
array bound 4-4, 6-1
array component 2-7
array constructor 3-4, 4-1, 6-17

array declaration 6-1
array element 4-1, 4-4, 6-4
    order, position 4-7
array expression 6-19, 7-2
array function 6-20
array operand 6-19, 7-2
array operation 6-19
array pointer 4-6, 5-1, 6-16
array processing 6-1
array reference 6-3
array section 4-1, 4-4, 6-5
array section of substrings 6-12
array specification 6-1
array subprogram 6-20
array variable 4-3
ASCII 1-1, 3-7, 7-8, 11-1, A-1
ASIN function 14-2, 14-9
assignment block 8-11
assignment interface block 8-7,
    13-21, 13-24
assignment statement 8-1
assignment subroutine 8-7, 13-9,
    13-20
assignment symbol 3-4
ASSIGNMENT(=) 8-7
ASSOCIATED function 14-2, 14-10
association 4-12, 13-29, 13-35
association status (of a pointer) 5-3
assumed length 9-13
assumed-shape array 4-6, 6-2, 13-34
assumed type parameter
    → assumed length
ATAN function 14-2, 14-10
ATAN2 function 14-2, 14-10

F

attribute 9-1, 9-2
automatic array 4-6, 6-12
automatic variable 4-11

BACKSPACE statement 11-39
BIT_SIZE function 14-2, 14-10
blank
   in input 12-5
   in source program 1-2
blank padding character 11-12,
      11-18, 11-26, 12-7
block 10-1
body of DO construct 10-10
branching 10-1
BTEST function14-2, 14-11

CALL statement 13-19
carriage control (character) 11-30
CASE construct 10-6
case expression 10-7, 10-8
case index 10-7
case selector 10-8
CASE statement 10-6, 10-8
case value 10-8
case-block 10-7
CEILING function 14-2, 14-11
CHAR function 14-2, 14-11
character assignment statement 8-4
character expression 7-12
character format specification 11-9,
    12-2
character length 2-4, 9-12, 9-13
character operator 7-12
character position 2-4, 4-4
character relational expression 7-8
character set
   ASCII 1-1, A-1
   F 1-1
CHARACTER statement 9-12
character string 2-4
character substring 4-4
character type 2-4

close (a file) 11-2, 11-33, 11-34
CLOSE statement 11-33
CMPLX function 14-2, 14-11
collating sequence 7-9, A-1
colon edit descriptor 12-8
comment 1-2
complex type 2-3
COMPLEX statement 9-11
component
   array component 2-7
   pointer component 2-7
   structure component 4-9
   type component 2-4, 2-6
component attribute 2-6
component definition 2-6
concatenation 7-12
condition
   end-of-file cond. 11-10, 11-11
   end-of-record condition 11-10
   error condition 11-10, 11-11
conformable 6-19, 7-1
CONJG function 14-2, 14-11
connection of a file (to a unit) 11-6
constant 4-2
constant array constructor 6-18
constant structure constructor 2-10
constant expression 7-22
construct 1-4
constructor 2-9, 3-4
CONTAINS statement 13-5
continuation (line) 1-1
control edit descriptor 12-4
control flow 10-1
control information list 11-8
control mask 8-12
COS function 14-3, 14-12
COSH function 14-3, 14-12
COUNT function 14-3, 14-12
count loop 10-10, 10-13
CSHIFT function 14-3, 14-12
current record 11-5
CYCLE statement 10-15

data edit descriptor 12-4
data entity, data object 4-1
data transfer statement 11-14
data type 2-1
DATE_AND_TIME subroutine
        14-6, 14-12
DEALLOCATE statement 5-5, 6-15
deallocation 5-4, 6-15, 6-17
declaration 9-1
declaration statement
        → specification statement
DEFAULT 10-6, 10-8
default type 2-1
defined
    assignment 8-6
    expression 7-13
    operator 7-14
defined 4-14
definition, definition status 4-14
delimiter 3-1
derived type 2-4, 13-7
derived type assignment stmt. 8-5
derived type definition 2-4
DIGITS function 14-3, 14-13
dimension (of an array) 6-1
DIMENSION attribute 6-1, 6-13, 9-3
direct access 11-3
direct input/output 11-3
DIRECT= specifier 11-35
disassociated 5-3
DO construct 10-10
DO statement 10-11
do-termination statement 10-11
DO variable 10-13
DOT_PRODUCT function 14-3, 14-13
dummy argument list 13-27
dummy array 4-7, 13-31, 13-33
dummy pointer 13-35
dummy subprogram 13-9, 13-21,
        13-28, 13-36

edit descriptor 12-1, 12-4
elemental (array) assignment 8-2
elemental function 14-1
elemental operation 6-20, 6-21, 7-1,
        7-20
elemental subprogram reference 14-7
else-block 10-1
ELSE statement 10-1
elsewhere-block 8-13
ELSEWHERE statement 8-12
elseif-block 10-1
ELSEIF statement 10-1
empty input list 11-13
empty record 11-14
empty value list 6-18
end of the record 11-21
end-of-file condition 11-10, 11-11
end-of-record condition 11-10
endfile record 11-2
ENDFILE statement 11-2, 11-40
ending position 4-4
END DO statement 10-11
END FUNCTION statement 13-14
END IF statement 10-1
END INTERFACE statement 7-15,
        8-7, 13-22, 13-23
endless loop 10-10, 10-14
END MODULE statement 13-3,
        13-5
END PROGRAM statement 13-2
END SELECT statement 10-6
END SUBROUTINE statement
        13-18
END TYPE statement 2-5
END WHERE statement 8-11
EOSHIFT function 14-3, 14-14
EPSILON function 14-3, 14-14
equality of derived types 2-5
equals, assignment symbol 3-4
equivalent expressions 7-3
error condition 11-11
ES edit descriptor 12-8

evaluation of expressions 7-20
executable construct 1-4
executable statement 1-4
execution control statement 10-1
execution part 13-2, 13-11, 13-13, 13-17
EXIST= specifier 11-35
EXIT statement 10-15
EXP function 14-3, 14-14
explicit initialization 9-10
explicit-shape array 4-7, 6-2
EXPONENT function 14-3, 14-14
exponent letter 12-8
exponent part 12-5, 12-8
exponent range 2-2, 14-26
expression 7-1
extent of a dimension 4-7, 6-6, 6-9
extended assignment 8-9
extended operator 7-17
external file 11-2
external subprogram 13-8, 13-20

F character set 1-1
F edit descriptor 12-10
F processor viii
field width 12-6
file 11-2
file attribute 11-2
file name 11-2
file position 11-5
file positioning statement 11-39
file status statement 11-31
FILE= specifier 11-31, 11-32, 11-35,
        11-38
FLOOR function 14-3, 14-15
FMT= specifier 11-9
form (of a file) 11-5
FORM= specifier 11-31, 11-35
format 12-1
format control 12-2
format item 12-1
format specification 12-1
formatted

file 11-5
    input/output 11-16
    record 11-1
FORMATTED= spec. 11-35
Fortran 90 D-1
FRACTION function 14-3, 14-15
fractional part 12-5, 12-8
function
    definition 13-8, 13-9, 13-11
    reference, invocation 13-14
function result 13-12
FUNCTION statement 13-11

generic interface block 7-15, 13-7,
        13-21
generic name 13-21, 13-23
global entity, global name 3-1
global scope 3-1, 13-8

high precedence (defined) expr. 7-18
host association 4-12
HUGE function 14-3, 14-15

I edit descriptor 12-10
IAND function 14-3, 14-15
IBCLR function 14-3, 14-15
IBITS function 14-3, 14-16
IBSET function 14-3, 14-16
ICHAR function 14-3, 14-16
IEOR function 14-3, 14-16
if-block 10-1
IF construct 10-1
IF THEN statement 10-1
implicit reference 7-15, 8-8
IMPLICIT statement 9-16
implied-DO 6-18
IN → INTENT attribute
INDEX function 14-3, 14-16
initial line 1-1
initial point (of a file) 11-5
initial value 9-3
initialization 9-9

initialization expression 7-24
INOUT → INTENT attribute
input argument 9-4, 13-27, 13-37
input field 12-5, 12-6
input list 11-13
input/output 11-1
input/output argument 9-4, 13-27, 13-37
input/output list 11-12, 12-2
input/output list item 11-12
input/output specifier 11-8
input/output statement 11-1, 11-7
input/output unit → unit
INQUIRE statement 11-34
inquiry (by unit, by file, by output list) 11-34
inquiry function 14-1, 14-7
instance (of a subprogram) 13-16, 13-20
INT function 14-3, 14-17
integer type 2-1
integer division 7-7
INTEGER statement 9-10
intended use of argument 9-4, 13-37
INTENT attribute 9-4, 13-37
interface → subprogram interface
interface block 13-21
interface definition 13-21, 13-23
INTERFACE statement 7-15, 8-7, 13-22, 13-23
internal file 11-2, 11-8, 11-25
internal input/output 11-24,
interpretation of expression 7-2, 7-4, 7-10, 7-13, 7-19
intrinsic
    assignment statement 8-1
    data type 2-1
    expression 7-2
    function 14-1
    operator 7-1
    subprogram 13-8, 13-25, 14-1
    subroutine 14-6
INTRINSIC statement 9-16

invalid operation 7-7
invocation 13-14
IOLENGTH= specifier 11-34
IOR function 14-3, 14-17
IOSTAT= specifier 11-10, 11-31, 11-32, 11-34, 11-35
ISHFT function 14-3, 14-17
ISHFTC function 14-3, 14-18
iteration count 10-12

keyword 3-1
keyword argument 13-29, 14-6
KIND dummy argument 14-7
KIND function 2-1, 14-4, 14-18
kind type parameter 2-1, 3-5, 3-6, 14-7
kind type param. conversion 7-5
KIND= 9-8

L edit descriptor 12-11
LBOUND function 14-4, 14-18
left tab limit 12-14
LEN function 14-4, 14-18
LEN= → length specification
length → character length
length specification 9-12
LEN_TRIM function 14-4, 14-19
lexical comparison 7-8
lexical token 1-2, 3-1
line
    printed line 11-30
    program line 1-1
list-directed formatting 11-20
list-directed input/output 11-20
literal constant 3-4
local entity, local name 3-1
local scope 3-1
LOG function 14-4, 14-19
logical type 2-3
logical assignment statement 8-3
logical expression 7-9
LOGICAL function 14-4, 14-19

logical operator 7-10
LOGICAL statement 9-11
LOG10 function 14-4, 14-19
loop
    DO construct 10-10
    implied-DO 6-17, 6-18
    nested loops 10-16
loop control 10-12
loop parameter 10-11
loop termination 10-13
low precedence (defined) expr. 7-18
lower array bound 6-1

main program 13-2
mask expression 8-12
MASK= argument 14-1
masked array assignment 8-11
MATMUL function 14-4, 14-19
MAX function 14-4, 14-20
MAXEXPONENT function 14-4, 14-20
maximum record length 11-32,
    11-33, 11-36
MAXLOC function 14-4, 14-20
MAXVAL function 14-4, 14-21
memory management 6-12
MERGE function 14-4, 14-21
MIN function 14-4, 14-21
MINEXPONENT function 14-4, 14-21
MINLOC function 14-4, 14-22
MINVAL function 14-4, 14-22
model number B-1
module 13-3
module function 13-9
module reference viii, 13-5
MODULE statement 13-3, 13-4
MODULE PROCEDURE statement
    7-15, 8-7, 13-24
module subprogram 13-5, 13-8
module subroutine 13-16
MODULO function 14-4, 14-22
multiple OPENS 11-33
MVBITS subroutine 14-6, 14-23

name 3-1
name association 4-12
NAME= specifier 11-35, 11-38
named constant 9-5
NAMED= specifier 11-35
NEAREST function 14-4, 14-23
NEQV → .NEQV.
nested constructs 10-5, 10-16
next record 11-5
NEXTREC= specifier 11-35
NINT function 14-4, 14-23
nonadvancing input/output 11-27
NONE → IMPLICIT statement
noneffective list items 11-12
nonexecutable statement 1-4
nonrepeatable edit descriptor 12-4
NOT function 14-4, 14-23
null value 11-23
nullification 5-6, 6-17
NULLIFY statement 5-6
NUMBER= specifier 11-35
numeric array 4-4
numeric assignment statement 8-2
numeric constant 3-4, 4-2
numeric editing 12-5
numeric expression 7-3
numeric operator 7-3
numeric relational expression 7-8
numeric variable 4-3

object → data object
ONLY → USE statement
only-list 13-6, 13-7
open (a file) 11-6
OPEN statement 11-31
OPENED= specifier 11-35
operand 7-1
operation 2-2, 2-3, 2-4, 7-7
OPERATOR
    → operator interface block
operator 3-4, 7-1

operator function 2-4, 7-14, 13-9

operator interface block 2-4, 7-15,
    13-21, 13-24

operator precedence 7-4, 7-5, 7-11,
    7-18

OPTIONAL attribute 9-4

optional dummy argument 13-29, 13-36

ordering of statements 1-5

OUT → INTENT attribute

output → input/output

output argument 9-4, 13-27, 13-37

output field 12-6

output list 11-14

output statement 11-1

overloaded subprogram name 13-24

PACK function 14-5, 14-24

padding character 11-12, 11-18,
    11-26, 12-7

PARAMETER attribute 9-5

parent object 4-2, 4-4, 4-9

parenthesized expression 7-19

pointer 4-1, 5-1, 8-2

pointer assignment stmt. 5-1, 8-9

pointer association 4-14, 5-1

POINTER attribute 9-6

pointer component 2-7

pointer function 13-12

pointer nullification 5-6, 6-17

pointer target 4-1, 5-2

position
    array element 4-8
    character substring 4-4
    file 11-5

POSITION= specifier 11-31, 11-33,
    11-35

positional argument 13-29

positioning statement 11-39

precedence 7-4, 7-5, 7-11, 7-18

preceding record 11-5

precision 2-2

PRECISION function 14-5, 14-24

preconnected file or unit 11-6, 11-7,
    11-8

present (argument) 13-36

PRESENT function 14-5, 14-24

PRINT statement 11-14

printing 11-30

private 2-8

PRIVATE attribute 2-5, 9-6, 9-14

private module 13-3

PRIVATE statement 9-14

procedure → subprogram

processor → F processor

PRODUCT function 14-5, 14-24

program 13-1

program library 13-8

program line 1-1

PROGRAM statement 13-2

program unit 13-1

pseudorandom number 14-6, 14-25

public 2-8, 13-3

PUBLIC attribute 2-5, 9-6, 9-15

PUBLIC statement 9-15

RADIX function 14-5, 14-25

railroad diagram E-1

RANDOM_NUMBER subroutine
    14-6, 14-25

RANDOM_SEED subroutine 14-6, 14-25

range 10-10

RANGE function 14-5, 14-26

rank 4-1, 4-7, 6-6

READ statement 11-14

READ= specifier 11-35

reading 11-1

READWRITE= specifier 11-35

real type 2-2

REAL function 14-5, 14-26

REAL statement 9-10

real-time clock 14-6, 14-30

REC= specifier 11-9

RECL= specifier 11-31, 11-33, 11-35

record 11-1, 11-25

F

record length 11-4
record number 11-3, 11-4
RECURSIVE 13-11, 13-17
recursive function 13-14, 13-16
recursive subroutine 13-18, 13-20
reference viii
    assignment subroutine 8-8
    function 13-14
    module 13-5
    operator function 7-15
    subroutine 13-19
relational expression 7-7
relational operator 7-8
repeat factor 11-23
REPEAT function 14-5, 14-26
repeat specification 12-1, 12-3
repeatable edit descriptor 12-4
representable character 2-4, 3-7
reserved word 3-2
restricted expression 7-25
RESHAPE function 14-5, 14-26
RESULT 13-11
result variable 13-11, 13-12
return 13-25
RETURN statement 13-26
reversion (of format control) 12-3
REWIND statement 11-39
RRSPACING function 14-5, 14-27

S edit descriptor 12-12
SAVE attribute 9-7
saved variable 4-15
scalar 4-1, 4-4
SCALE function 14-5, 14-27
SCAN function 14-5, 14-27
scope 3-1
scoping unit 3-1
section subscript 6-6
SELECTED_INT_KIND function
    14-5, 14-28
SELECTED_REAL_KIND function
    14-5, 14-28

selector 10-8
SELECT CASE statement 10-6
separator
    format specification 12-1
    internal input 11-26
    list-directed input 11-21
sequential access 11-3
SEQUENTIAL= specifier 11-35
set of values
SET_EXPONENT function 14-5,
    14-28
shape 4-7, 6-6
SHAPE function 14-5, 14-28
side effect 6-4, 7-20, 8-14, 11-8
sign control edit descriptor 12-5,
    12-12
SIGN function 14-5, 14-29
significand 12-9
simple executable statement 1-4
SIN function 14-5, 14-29
SINH function 14-5, 14-29
SIZE function 14-5, 14-29
size of an array 4-7, 6-6
SIZE= specifier 11-11
slash edit descriptor 12-13
source form 1-1
SP edit descriptor 12-12
SPACING function 14-5, 14-29
special character 1-1
special expression 7-22
special name 3-4
specific interface block 13-20, 13-21,
    13-22
specific name 13-11, 13-18, 13-21
specification expression 7-25
specification part 13-2, 13-3, 13-4,
    13-11, 13-17
specification statement 9-1
SPREAD function 14-5, 14-29
SQRT function 14-5, 14-30
SS edit descriptor 12-12
standard unit 11-7

starting position 4-4
STAT= 5-2, 5-5, 6-14, 6-15
statement keyword 3-1
statement ordering 1-5
status variable 5-2, 5-5, 6-14, 6-15
status
    allocation status 6-14
    association status (of a pointer)
        5-3
    definition status 4-14
    input/output status information
        11-10, 11-31, 11-32, 11-34,
        11-35
STATUS= specifier 11-31, 11-33, 11-34
STOP statement 10-16
stride 6-8, 6-18, 10-11
structure component 4-9, 6-5
structure constant 2-10
structure constructor 2-9
structure object 2-9, 4-1, 13-7
subobject 4-2, 4-4
subprogram 13-1, 13-8
subprogram definition 13-11, 13-17
subprogram interface 13-9
subroutine 13-8, 13-16
    definition 13-17
    reference, invocation 13-19
SUBROUTINE statement 13-17
subscript expression 6-6
subscript list 6-4
subscript value sequence 6-6, 6-8
subscript-triplet 6-8
substring 4-4
SUM function 14-6, 14-30
syntax chart E-1
SYSTEM_CLOCK subroutine 14-6,
    14-30

T edit descriptor 12-13
tabulator edit descriptor 12-13
TAN function 14-6, 14-31
TANH function 14-6, 14-31

target 4-1
TARGET attribute 9-8
temporary file 11-33
terminal point (of a file) 11-5
THEN → IF THEN statement
TINY function 14-6, 14-31
TL edit descriptor 12-13
token → lexical token
TR edit descriptor 12-13
transfer of control 10-1
transformational function 14-1
TRANSPOSE function 14-6, 14-31
TRIM function 14-6, 14-31
type 2-1
type component definition 2-6
type concept 2-1
type conversion 7-5
type declaration statement 9-1, 9-8
type definition 2-4
type parameter →
        KIND type parameter
type specification 9-8
TYPE statement
    type declaration 9-13
    type definition 2-5

UBOUND function 14-6, 14-32
undefined 4-14, 5-3
unformatted
    file 11-5
    input/output 11-18
    record 11-1
UNFORMATTED= specifier 11-35
unit 11-6
unit number 11-6, 11-8, 11-9
UNIT= specifier 11-8, 11-34, 11-35
UNPACK function 14-6, 14-32
upper array bound 6-1
USE association 4-12, 13-6
USE statement 13-5

value list 6-17

F

value separator 11-21
variable 4-3
variable (dummy) array 4-7, 13-34
variable format specification 12-2
vector subscript 6-11
VERIFY function 14-6, 14-32
visible 2-8, 9-6

where-block 8-13
WHERE construct 8-11
WHERE construct statement 8-11
whole array 6-3
WRITE statement 11-14
WRITE= specifier 11-35
writing 11-1

zero-length character string 7-21
zero-size array 6-2, 7-21

.AND.   7-10, 7-18
.EQV.   7-10, 7-18
.FALSE.   3-6
.NEQV.   7-10, 7-18
.NOT.   7-10, 7-18
.OR.   7-10, 7-18
.TRUE.   3-6
==   7-8, 7-18
>=   7-8, 7-18
>   7-8, 7-18
<=   7-8, 7-18
<   7-8, 7-18
/=   7-8, 7-18
=>   8-9